Library of
Davidson College

TERRESTRIAL SLUGS

Biological Sciences

Editor
PROFESSOR A. J. CAIN
MA, D.PHIL
Professor of Zoology
in the University of Liverpool

TERRESTRIAL SLUGS

N. W. Runham, Ph.D
Lecturer in Zoology,
University College of North Wales, Bangor

&

P. J. Hunter, Ph.D
International Scientific Relations Division,
Department of Education and Science, London

HUTCHINSON UNIVERSITY LIBRARY
LONDON

HUTCHINSON & CO (*Publishers*) LTD
178–202 Great Portland Street, London W1

London Melbourne Sydney
Auckland Johannesburg Cape Town
and agencies throughout the world

First published 1970

The paperback edition of this book is sold
subject to the condition that it shall not, by
way of trade or otherwise, be lent, re-sold,
hired out, or otherwise circulated without
the publisher's prior consent, in any form of
binding or cover other than that in which
it is published and without a similar
condition including this condition being
imposed on the subsequent purchaser

© N. W. Runham and P. J. Hunter 1970

*This book has been set in Times type, printed in Great Britain
on smooth wove paper by Anchor Press, and
bound by Wm. Brendon, both of Tiptree, Essex*

ISBN 0 09 105670 5 (cased)
ISBN 0 09 105671 3 (paper)

CONTENTS

	Preface	7
1	General features of slugs	9
2	The classification of slugs	21
3	Food, feeding and digestion	37
4	Metabolism, respiration, breathing, circulation, blood, water relations and excretion	59
5	Reproduction, development, growth and genetics	83
6	Locomotion, mucus, sensory structures, nervous system, endocrinology	98
7	Ecology	116
8	Slugs as pests	138
	References	155
	Index	177

PREFACE

At first sight slugs appear to be improbable land animals, as they have a high water content but no impermeable cuticle; yet they are extremely successful. Their success is due to their considerable adaptability.

In this book we have attempted to survey the physiology, behaviour, ecology and economic importance of slugs. It was written firstly to provide information for those people seeking to control these important agricultural and horticultural pests. Current trends in agriculture seem likely to make them even more serious pests in the future. It is therefore essential to discover better methods of control. The mode of action of the current molluscicides is largely unknown and therefore their further development is a matter of chance. Perhaps a better understanding of slug physiology, behaviour and ecology will lead to the design of more effective chemical or biological control methods.

Secondly, it was hoped that the book would promote the use of slugs in schools and universities as material for research and teaching, where the demand is for animals easily collected in large numbers. Apart from their pest status slugs have an intrinsic biological interest. It is now possible to obtain suitable keys to identify the local species in many parts of the world but it is difficult to find information on their general biology.

Most of the work on land molluscs has concentrated on species of the snail *Helix* and slugs get little mention in general reviews. It therefore seemed that it would be of value to present as full a picture as possible of the life of slugs. In some fields of study where little work has been carried out with slugs, information has had to be included from studies on related molluscs. It is regrettable that little work has been carried out on tropical slugs, so that most of the information utilised here had to be taken from studies on North European species. There is a need for further information at all levels from the simplest natural history to the most sophisticated physiology and ecology, and in all countries. Perhaps it is in the field of taxonomy that work is most urgently required, as this information is basic for all other work.

The views expressed in this book are the joint responsibility of the authors but Chapters 1, 3, 4, 5 and 6 have been largely written by NWR, and Chapters 2, 7 and 8 by PJH. The physiological chapters owe much to the stimulus, information and criticism provided by a series of PhD students (B. J. Smith, C. J. Bayne, G. Walker, T. G. Bailey, A. A. Laryea and J. G. Garner). Professor A. Milne and Mr J. R. Vernon made valuable comments on the ecology chapters. We are grateful to Professor A. J. Cain, Dr D. L. Gunn, Mr H. Gould, Dr T. B. Reynoldson, Mr A. E. Ellis, Dr V. Fretter, and Mr J. Peake for reading and criticising the manuscript. Without the considerable help provided by the Science Library, University College of North Wales, Bangor, the Agricultural Advisory Service and the Balfour Libraries, Cambridge, the book could not have been written in the time available. The drawings were executed by NWR and Mrs Marilyn Runham, and the photographs were taken by Mr T. G. Bailey and Mr B. V. Symonds. We are grateful to the Controller of Her Majesty's Stationery Office for permission to use Figures 53 and 55.

I

GENERAL FEATURES OF SLUGS

INTRODUCTION

A superficial examination of the head and foot of a slug is sufficient to reveal that they are closely related to snails, and they belong to that very successful group of animals known as molluscs. A study of the detailed anatomy of slugs and a survey of the molluscs show that the slug form has evolved several times in this group of animals both in the sea and on land. Slugs are therefore not a natural group of closely related animals; the term 'slug' refers only to a type of body shape. To understand the anatomy of these animals it is first necessary to survey the phylum Mollusca. There are several excellent reviews on this group (Hyman 1967, Franc 1968, Morton 1958) and only a very brief summary will be given here.

Characteristically, molluscs have a body divided into two; a visceral mass containing most of the organs and a combined head and foot. A shell, secreted by a specialised area of the body wall called the mantle, covers the visceral mass. There is no internal supporting skeleton and the skin secretes abundant mucus so that the body is soft and slimy to the touch. Respiration is effected by gills in a pallial or mantle cavity. Internally the body is not subdivided by segmentation and the body cavity is blood-filled (haemocoel). The feeding organ is a horny rasping strap, termed the radula, which is peculiar to the molluscs. The nervous system is ganglionated.

The molluscs are a very varied group; they have invaded most habitats, and in number of species are exceeded only by the arthropods. Molluscs can be subdivided fairly readily into seven groups, although there is some disagreement as to the systematic rank of these—here they will all be referred to as classes.

CLASS MONOPLACOPHORA
Adults have a simple cap-like shell and some of the organs are serially repeated. Rare and primitive deep sea forms e.g., *Neopilina*.

CLASS POLYPLACOPHORA
Eight dorsal shell plates. Serial repetition of the gills. Marine e.g., *Lepidochitona* (coat of mail shell or chiton).

CLASS APLACOPHORA
Worm-like, shell represented by spicules embedded in the body wall. Marine e.g., *Chaetoderma*.

CLASS GASTROPODA
Visceral mass characteristically rotated through 180° relative to the foot (this rotation is termed torsion). At some stage a helical shell is present. Marine, fresh water and terrestrial e.g., *Littorina* (winkle), *Buccinum* (whelk), *Helix* (snail), *Arion* (slug).

CLASS SCAPHOPODA
Tube-like shell. Marine e.g., *Dentalium* (elephant tusk shell).

CLASS BIVALVIA
Two shell valves united by a ligament. Radula absent. Marine and freshwater e.g., *Ostrea* (oyster), *Anodonta* (swan mussel).

CLASS CEPHALOPODA
Very highly advanced forms; swim by jet propulsion; arms surround mouth. Marine e.g., *Loligo* (squid), *Octopus* (octopus).

The class Gastropoda can be subdivided into three subclasses and two of these contain large groups of slugs.

SUBCLASS PROSOBRANCHIA
Gills present; torsion clearly evident. Marine, freshwater, and terrestrial e.g., *Littorina* (winkle).

SUBCLASS OPISTHOBRANCHIA
Strong tendency for reduction or loss of shell. Torsion wholly or partially reversed. Marine and terrestrial, e.g., *Aplysia* (sea hare), *Doris* (sea slug).

SUBCLASS PULMONATA
Gills absent, being replaced by a 'lung'. Great concentration of nerve ganglia so that torsion, usually most readily seen in the

General features of slugs

nervous system, is difficult to visualise. Freshwater and terrestrial e.g., *Helix* (snail), *Arion* (slug).

Within the opisthobranchs there are many slug-like forms and the vast majority of these are marine. These are the extremely beautiful sea slugs or nudibranchs (e.g., *Doris, Eolis*). In addition there are a few forms, the Onchidoidea, which have invaded land, although they are usually found near water. This small group of slugs is often included with the pulmonates, but, as they appear much more closely related anatomically to the opisthobranchs, many modern authors include them in this latter group.

The slugs which are considered in detail here are all pulmonates. This group, which contains the vast majority of terrestrial molluscs, is divided into three orders:

Order Basommatophora
Nearly all aquatic; a single pair of tentacles with eyes at their bases e.g., *Lymnaea* (pond snail), *Planorbis* (ramshorn).

Order Systellommatophora
A small group of tropical slugs; terrestrial; two pairs of tentacles with eyes at the tips of the posterior pair; tentacles contractile.

Order Stylommatophora
Terrestrial snails and slugs; two pairs of tentacles, the posterior pair bearing eyes at their tips; tentacles retractile.

We can see that even at the level of subclass and order there are groups of animals which are only distantly related but yet are known as slugs. The most obvious and striking characteristic of slugs is the strong reduction or absence of a shell. Perhaps even more significantly, the organs which in other molluscs are housed in the visceral mass protected by the shell, are here incorporated into the head-foot. The slug shape is therefore streamlined and in some cases almost worm-like (Fig. 1). As this form has been adopted independently by several groups of molluscs, presumably by parallel evolution, there must surely be important advantages associated with it. The availability of calcium restricts the distribution of terrestrial shell-bearing molluscs as the shell is mainly hardened by calcium salts. The reduction or loss of the shell has enabled slugs to tolerate far lower levels of calcium in their environment than most snails. A few snails

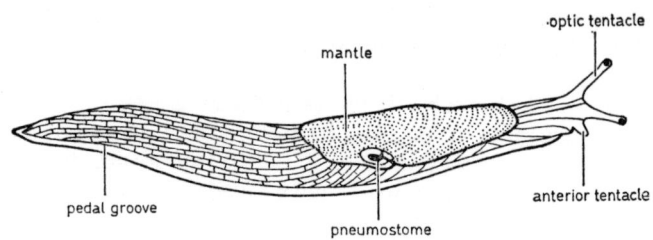

Fig. 1 *Agriolimax reticulatus*, the grey field slug, external features.

have reduced their calcium requirements by producing a horny shell but these snails tend to be small. By the incorporation of the visceral mass in the muscular head-foot the animal becomes more streamlined and so can move through crevices or squeeze its body through small holes. Some snails are found in crevices but these are very small in size. Slugs, even those which attain a large size, are therefore often able to live in habitats unsuitable for snails.

Very little is known of the evolution of the slugs because the soft parts are not preserved in the fossil record. Sequences can be constructed from related living snails to illustrate the gradual loss of shell and incorporation of the visceral mass into the head-foot. Slugs and snails have a common developmental stage in which the cap-shaped shell is contained in an internal sac with most of the organs developing near it. In the snails the shell later becomes external and as it grows it becomes helically coiled. The organs grow within the enclosed visceral mass. Arionid and limacid slugs retain the internal shell and the organs grow instead into the cavity of the foot. It is therefore possible to conjecture that the slugs arose by a process of neoteny (the retention of early developmental characters by the adult).

EXTERNAL ANATOMY

In slugs the typical spiral gastropod shell has become reduced or is completely absent. Where present it is of the same basic spiral pattern but has become ear-shaped or flattened. In a few forms it is external (e.g., *Testacella,* Figs. 2 and 10) while in most it is completely internal (e.g., *Agriolimax*), being contained within a shell sac.

As the viscera are contained within the head-foot, the outline of the animal is smooth and streamlined. The head bears two pairs of

Fig. 2 *Testacella maugei* (after Taylor 1907). Note the external shell.

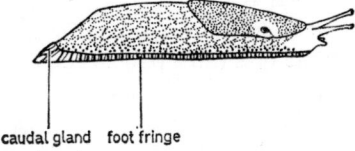

Fig. 3 *Arion ater*.

Fig. 4 *Veronicella moreleti* (after Franc 1968).

tentacles (Fig. 6), except in one group, the Athoracophoridae, where only a single pair is present. There is an eye at the tip of each of the larger posterior or optic tentacles and this is visible as a small black dot. Apart from the absence of the eye the anterior tentacles are similar, except in the Veronicellidae where they are bilobed. Most slugs can retract their tentacles completely by inversion

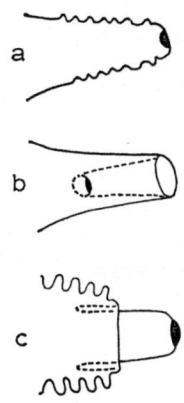

Fig. 5 Methods of tentacle retraction (after Burch 1968).
   a. Contraction, e.g. *Vaginulus*.
   b. Inversion, e.g. *Agriolimax*.
   c. Contraction and invagination, e.g. *Aneita*.

(Fig. 5b). The Systellommatophora can however only contract their tentacles (Fig. 5a). In the Athoracophoridae (e.g., *Pseudaneitea*) the smooth rod-like tip of the tentacles (Fig. 5c) can be drawn back into the base without inversion, and the base then contracts (Burch 1968).

At the tip of the head there is a simple mouth surrounded by a complex arrangement of lips and flanked by the mouth lobes which may represent remnants of a third pair of tentacles (Fig. 6). The head

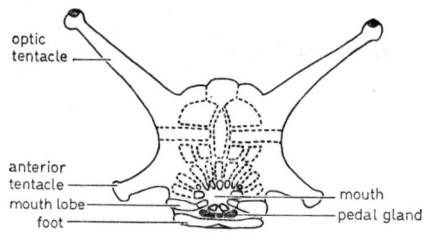

Fig. 6 *Agriolimax reticulatus*, front view of the head (after Walker 1969).

is not clearly separated from the body but there is usually a very extensible thin-wall neck region. Over the general surface of the body the skin is ciliated, and can bear tubercles which may be large and prominent (e.g. *Arion ater* Figs. 3 and 12) or so flat that the surface of the animal appears almost smooth (*Agriolimax caruanae*). A pattern of grooves may be present between the tubercles, but in the Athoracophoridae there is a complex arrangement of slime furrows and papillae (Fig. 20). In some species there is a line of tubercles and/or a shaping at the rear of the body to form a keel on the back (*Milax* spp Fig. 14).

That area of the body surface covered by the shell in snails is termed the mantle and it usually has a thickened edge or collar. In slugs this region is considerably modified and forms a characteristic area of the body surface. The mantle in *Agriolimax* occupies about a quarter of the dorsal body surface and is finely grooved (Figs. 1 and 13). Anteriorly the mantle is drawn out into a large mantle flap which covers the head and neck except when they are fully extended. On the right of the mantle there is a large opening to the lung, the pneumostome, and an associated groove into which open the rectum and ureter (Fig. 33). Around the pneumostome there is a specialised glandular area visible externally as a lightly pigmented ring. There is considerable variation in the texture of the mantle surface, from the fine grooving found in members of the Limacidae, to the coarse

General features of slugs

warty appearance of the Veronicellidae. Athorocophoridae have a minute mantle area distinguishable only with difficulty (Fig. 20), while in the Veronicellidae (Fig. 4) it constitutes the whole dorsal surface of the body.

The sole of the foot is usually the same width as the body. In the Veronicellidae however it is very narrow and surrounded by a deep groove. These animals have the mantle hanging down over the sides of the foot, the underside of this mantle being termed the hyponotum, its edge the perinotum, and its dorsal surface the notum. If slugs are viewed from beneath while crawling on glass alternate light and dark transverse bands can be seen passing anteriorly along the surface of the foot (Fig. 42). Some groups of slugs have the foot sole undivided (e.g., *Philomycus*) while in others it is longitudially tripartite and the locomotory waves involve only the central part (*Agriolimax*). Typically the surface of the foot extends a short distance up the sides of the body (aulacopod condition) and it is delimited from the general body surface by the pedal groove and sometimes also by the suprapedal groove. This part of the foot may form a well defined foot fringe (e.g., *Arion ater*, Figs. 3 and 12). Opening out over the front of the foot from beneath the head is a large funnel-shaped opening, the pedal gland, which secretes the mucus on which the animal crawls. Mucus is also secreted by the other body surfaces and may be characteristically coloured (e.g., the yellow mucus of *Arion hortensis*). A very thick mucus is produced in some species by a large gland at the posterior end of the body, the caudal gland.

INTERNAL ANATOMY

The mouth leads into a buccal cavity which receives the secretions of the salivary glands and contains the radula and jaw (Figs. 7, 21 and 26). The jaw is a hardened transverse plate, or in the Veronicellidae a group of plates, in the roof of the buccal cavity. Projecting into this cavity from the floor is the odontophore, consisting of the radula, the radular gland which secretes the radula, the supporting odontophore cartilages, and associated musculature. The radula is a flexible strap bearing hard pointed processes or 'teeth' which functions as a rasp, or in *Testacella* as a grasping organ. Each radular tooth (Figs. 25 and 22) consists of one or more cusps mounted on a base, the large central cusp being termed the mesocone, and the outer ones ecto- or endocones depending on their position. Several types of teeth may be present. The central or rachidian tooth is symmetrical

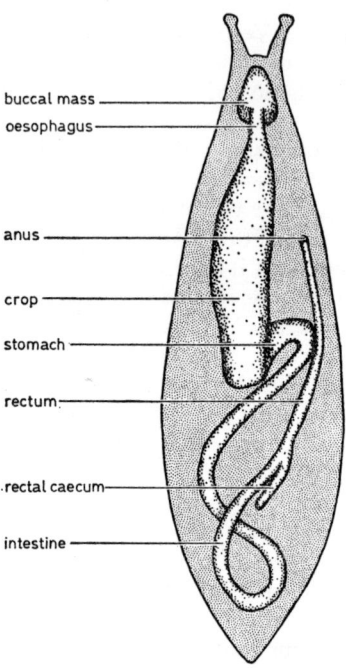

Fig. 7 *Agriolimax reticulatus*. A diagram showing the position of the digestive tract within the body.

while the lateral teeth are asymmetrical. The lateral teeth grade into the usually much simpler marginal teeth.

Food passes from the buccal cavity by way of a short oesophagus into a voluminous crop and thence to the stomach (Figs. 7 and 26). The stomach is small and simple and from it two small ducts pass to the very large digestive glands. Faeces form at the posterior end of the stomach and pass through the coiled intestine to the thin-walled rectum. This opens at the anus near the pneumostome, except in the Athoracophoridae where it is displaced laterally. In some Limacidae a small caecum is found at the junction of the intestine and the rectum. The external opening of the ureter is also associated with the pneumostome. As there is considerable variation in the morphology of the excretory system it is an important systematic character. The kidney is sac-like with internal folds, and the semi-solid excreta are transported through the ureter. The Stylommatophora can be sub-

General features of slugs

divided into four suborders by the arrangement of the kidney and ureter:

Orthurethra: ureter completely separate from the rectum and reflexed near its tip;

Mesurethra: no ureter, kidney opens to the exterior directly;

Heterurethra: while in all the other groups the long axis of the kidney is parallel to the antero-posterior axis of the mantle, in this group it is transverse; the ureter runs along the anterior border of the kidney and then alongside the rectum;

Sigmurethra: the ureter is reflexed back along the lateral edge of the kidney, then turns sharply forwards to run alongside the rectum (Fig. 32).

The pneumostome leads into the sac-like lung (Fig. 8c and d) and part of the wall of this is highly folded and vascularised to form the respiratory epithelium. In Athoracophoridae the lung is small with many fine tubules leading off it, which were thought to resemble the tracheae of insects, hence the alternative name for this group—the Tracheopulmonata. The Veronicellidae lack a mantle cavity.

The pericardium, containing the heart, is usually located in the floor of the lung (Figs. 8b and 32). The heart consists of a thin-walled auricle and a very muscular ventricle. Between the pericardium and the kidney runs a small highly ciliated renopericardial duct. The aorta arises from the ventricle and divides into anterior and posterior aortae. These aortae divide into an extensive network of arteries supplying the organs (Fig. 30), which in turn break up into smaller arteries, arterioles, and very fine vessels termed capillaries. The latter open directly into the blood sinuses or the haemocoel. Blood from the various sinuses passes to the auricle via the kidney and lung. The organs lie in the blood-filled body cavity or haemocoel which is compartmented by a number of septa.

Slugs are all hermaphrodite. The gonad contains a mixture of gametes (Fig. 43) but usually the sperm are mature slightly before the ova (protandry). From the gonad (Figs. 9 and 35) the hermaphrodite duct conveys the gametes to the complex glandular reproductive tract. At the junction an albumen gland is present. In Systellommatophora there are separate male and female ducts (Fig. 36), but in the majority of slugs these are fused to give a common duct (Fig. 35). The external opening is usually single (monaulic) but in the Systellommatophora there are separate openings (diaulic). The external genital openings are usually difficult to see except during courtship. The morphology of the tract is extensively used in systematics for species

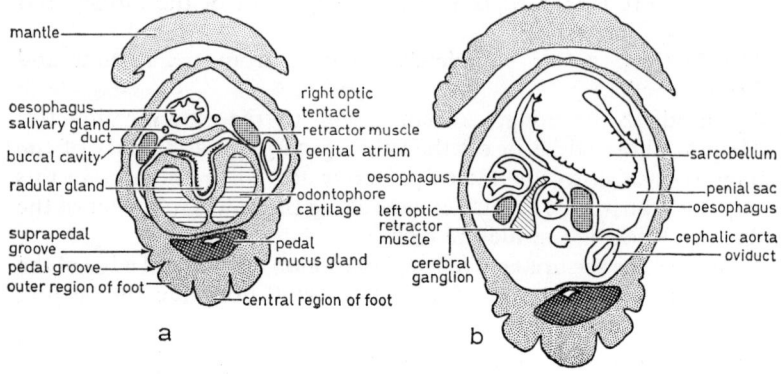
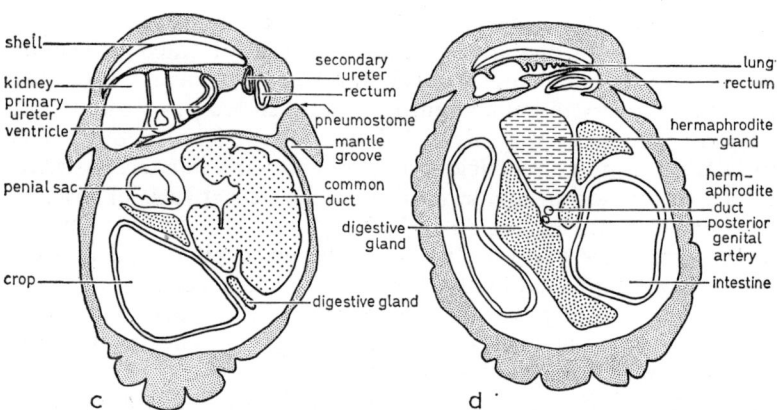

Fig. 8 Transverse sections through the body of *Agriolimax reticulatus*.
 a. Anterior head region.
 b. Posterior head region.
 c. Pneumostome region.
 d. Posterior mantle region.

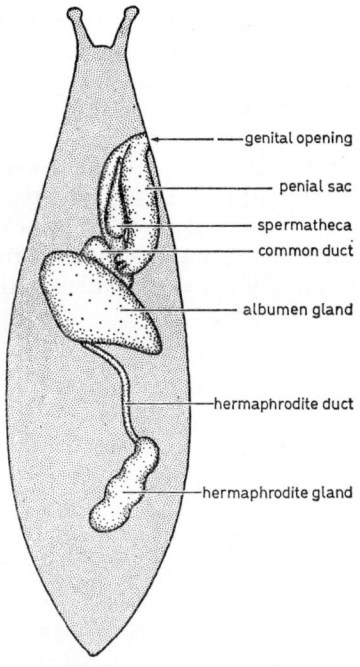

Fig. 9 *Agriolimax reticulatus*. A diagram showing the position of the reproductive system within the body.

differentiation. Courtship is complex and protracted, leading to the eversion of some of the reproductive organs and then coitus. In some groups (e.g., *Arion*) the sperm are packaged in spermatophores (Fig. 38), which are exchanged at coitus, while in other groups these are absent (e.g. *Limax*). Various stimulatory appendages are present: in some species of *Arion* there is an erectile ligule on the wall of the oviduct; in *Agriolimax* there is an evertile sarcobelum on the wall of the penis sac (Fig. 41); and in *Vaginulus* there is a dart sac with a calcified dart. The zygote is coated with a nutritive layer of albumen from the albumen gland and then surrounded by various protective coats (Fig. 39) on its passage through the female parts of the reproductive tract to form the egg. The eggs are large and are usually laid in batches.

A nerve ring consisting of paired cerebral, pleural, parietal and pedal ganglia together with a single visceral ganglion is situated around the oesophagus (Fig. 48). These ganglia are extensively fused so that it is usually difficult to differentiate them externally. A well developed network of nerves supplies the organs (Fig. 49) and on the surface of these they form very extensive plexuses.

The muscular system is complex and consists of three main divisions. In the foot sole there is a thick meshwork of muscles responsible for locomotion; in the body wall there is another meshwork of muscles bringing about postural movements of the body; and there is a set of internal muscles inserted into the floor of the pulmonary sac, or around its edge, which retract the penis, the tentacles and the buccal mass.

2

THE CLASSIFICATION OF SLUGS

Slugs are notoriously difficult to describe and identify because of the softness of their bodies and the extreme variation in colour and size. The shell, by which most of their molluscan relatives can be identified, is usually internal or absent. Furthermore, slugs arising from widely divergent origins have, by parallel evolution, assumed a remarkable similarity in their general features.

Difficulties in describing and identifying slugs have given rise to many misconceptions about the geographical distribution of certain species. For example, *Agriolimax reticulatus* was for many years wrongly identified as *Agriolimax agrestis* in many countries. Moreover, some taxonomists and systematists have worked from only a small number of preserved specimens, whose shape and colour have become distorted by the preservative. Thus, many specimens have been allocated to the wrong species or wrongly described as 'new'. There has been very little effort given to the comparison of specimens from different parts of the world. In this account, the geographical distribution of species has been presented in a deliberately general form. It can be assumed that some species do not occur in places where they have been recorded and, more likely, that they do occur in many areas where they have not yet been found.

There is no authoritative account of the classification of slugs. In particular, the arrangement of families containing slugs into larger groups (super-families or sub-orders) has not been fully established. Several systems have been suggested and it is not possible for the non-specialist to decide which is the best—we will therefore not refer to super-families. Sub-family groupings are also not well defined and we will refer to sub-families only where this appears to be of particular value.

The scheme of classification used here (Table 1) is based largely on those of Burch (1962) and Franc (1968).

TABLE I
CLASSIFICATION OF THE PULMONATA

Order	Sub-order	Family
Basommatophora		(No slugs)
Systellommatophora		Rathousidae
		Veronicellidae
Stylommatophora	Orthurethra	10 families (No slugs)
	Mesurethra	6 families (No slugs)
	Heterurethra	Athoracophoridae
		2 other families including the succinid snails
	Sigmurethra	Arionidae
		Philomycidae
		Limacidae
		Testacellidae
		25 other families including the Helicidae. In several families e.g., Endodontidae and Helicarionidae many species have much reduced shells

The main features used for the description and identification of slugs are: the position and shape of the radular teeth; the shape of the jaw; the shape, size and position of various parts of the reproductive system; and the position of major muscle groups. For the identification of particular species, reference should be made to one of the standard descriptive works: Quick (1960) for British species,

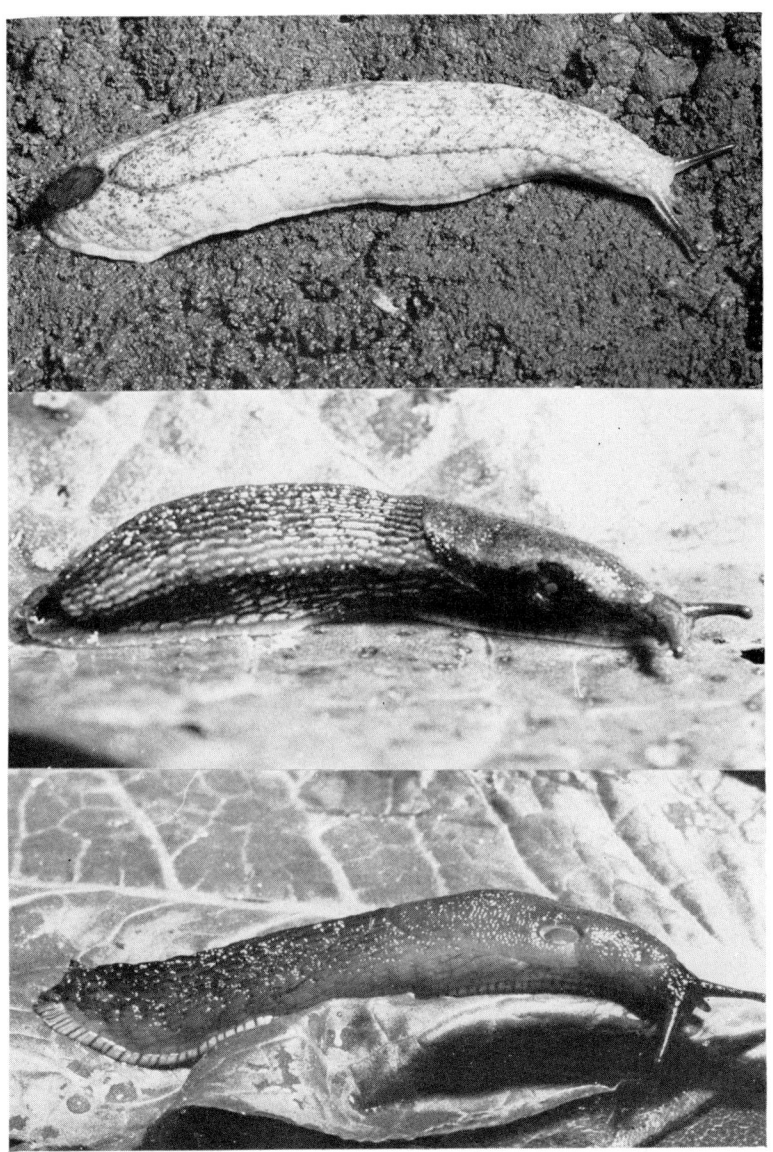

Fig. 10 (ABOVE) *Testacella scutulum.*
Fig. 11 (CENTRE) *Arion hortensis.*
Fig. 12 (BELOW) *Arion ater.*

Fig. 13 (ABOVE) *Agriolimax reticulatus*.
Fig. 14 (CENTRE) *Milax budapestensis*.
Fig. 15 (BELOW) *Limax maximus*.

Pilsbry (1948) for North American species, Faune de France (Germain 1930) for European Continental species, Likharev and Rammel'meier (1962) for East European species, Burton (1962, 1963a,b) for the Athoracophoridae. There are also good keys for the identification of slugs in certain areas, for example Ellis (1969) and Quick (1949) for British slugs and Burch (1962) for Eastern American slugs.

THE RANGE OF FORM AND HABITS OF SLUG FAMILIES

This account includes a wide ranging discussion of slug families paying most attention to those species mentioned repeatedly later.

Family TESTACELLIDAE

The most noticeable characteristics of these slugs (Figs. 2 and 10) is the small ear-shaped shell at the rear end of the body protecting the heart, kidney and mantle cavity. A pair of grooves run along the side of the body between the front of the mantle and the back of the head. The upper tentacles are more pointed than those of other slugs and there is no caudal mucus gland.

These slugs are carnivorous and eat earthworms and other soil invertebrates, including slugs. The alimentary canal is specially adapted to their carnivorous feeding habits, but there is also a major reorganization apparently unrelated to the diet. Torsion, or the rotation of the viscera so that the genital opening and anus face forward, has been reversed. The normal, backwardly directed, loop of the intestine has been obliterated and the rectum runs backwards to the anus which lies inside the pulmonary cavity. The jaw is absent. Sharp needle-like radular teeth (Fig. 25b) are used to impale the prey. The buccal mass is much enlarged and muscular but the stomach is feebly developed.

These slugs are mostly found on cultivated ground. They remain underground or under stones or leaf litter during the day, and in the winter they hibernate in an underground cavity. There is only one genus, *Testacella*,[1] containing three important species: *Testacella maugei*, **Férussac 1819**, found in the west of Europe; *Testacella haliotidea*, **Draparnaud 1801**, ranging from the North African coast to Scotland and from the Atlantic Islands to the Balkans; and *Testacella scutulum*, **Sowerby 1820**, with an ill-defined distribution in Europe.

---

[1] The authority for each species is given in bold type when it is first mentioned.

Family ARIONIDAE

These slugs are descended from an old and widespread family of primitive land snails with a flat coiled shell (Endodontidae). The North American genus *Binneya* retains an external spiral shell, which in the European genus *Geomalacus* is reduced to an oval internal shell, and is further reduced in *Arion* to a mass of granules. The mantle of these slugs is located near the front of the body and the respiratory opening is on the right of the front of the mantle. Usually the rear end of the body is rounded and there is no keel. There is often, however, a prominent row of tubercles between the posterior end of the mantle and the rear end of the body, which, in young *Arion fasciatus* for example, has the appearance of a keel. The foot is wider than the rest of the body, the lateral edges being termed the foot fringe. The upper tentacles are not sharp and tend to have rounded ends.

The pericardium, surrounding the heart, is enclosed by the kidney. The jaw is ribbed. The central radular teeth are tricuspid and the lateral teeth bicuspid. The marginal teeth have short, wide basal plates with long mesocones, and ectocones may or may not be present. Arionid slugs live in a wide variety of habitats, usually in areas of relatively high humidity, throughout the temperate regions of the world. There are three sub-families. The Arioninae do not possess a penis or vagina and the cephalic retractor muscles are widely separated at their origin from the posterior border of the lung floor. The apex of the stomach lies behind the posterior loop of the intestine. This sub-family is common throughout temperate regions. It contains the generally distributed genus *Arion*, and the very localised genus *Geomalacus*, which (with the exception of the spotted Kerry slug, *Geomalacus maculosus*, **Allman 1843**, from Ireland) is restricted to the Iberian peninsula.

There are several species. In some species a ligule (stimulatory appendage) is present on the wall of the oviduct (*Arion lusitanicus, A. subfuscus,* and *A. hortensis*); in others it is present on the wall of the genital atrium (*A. ater*); in others it is lacking (*A. fasciatus*). Three species are very common throughout all temperate regions of the world, often becoming pests of agricultural or horticultural crops.

Arion ater ater, (**Linnaeus 1758**), is a large black slug and is probably the species most readily noticed by the casual observer. The extended length of the body can reach 15 cm and 2 cm in breadth. The skin is tough and marked by longitudinal rows of coarse tubercles (Figs. 3 and 12). The foot is light in colour and a wide fringe is traversed by

The classification of slugs

alternating thin and thick stripes. The mucus is copious, very sticky and white. The young are straw coloured and usually have no longitudinal dark bands (as in *Arion ater rufus*, (**Linnaeus 1758**)). As the young grow, a dark area on the dorsal side gradually extends downwards towards the foot. Very occasionally the slug remains completely white but these albino individuals are rare. A peculiar habit of these slugs is to contract into a hemisphere when disturbed and rock from side to side at about two-second intervals.

This species is herbivorous, feeding on decaying vegetable matter and growing plants but also eats animal material or faeces. It is often common in grass fields or hedgerows, and is more easily noticed than smaller slugs so that even when recorded in high numbers it may not be the commonest species present. It is found in gardens, hedgerows, agricultural land, moors and bogs, and often in colder regions. This species is widely distributed throughout Europe, Russia and North America. In northern Europe, it has an annual life cycle. It lays batches of large (usually 5×4 mm) opaque eggs in late summer or autumn. The first cluster contains up to 150 eggs, and two or three smaller clutches are laid later. The eggs probably remain unhatched over winter and the young emerge in spring. These grow quickly to mature in late summer.

Arion hortensis, **Férussac 1819**, is probably the commonest *Arion* species (Fig.11). It is small and slender, without a foot fringe. The colour varies from dark grey to black, while some are dark brown. The foot is yellow or orange and the slime is also an orange/yellow colour. The extended length of adults is usually about 2–4 cm. Juveniles are bluish grey when newly hatched.

These slugs have an annual life cycle in northern Europe. They breed in mid-summer and the young grow during the autumn, winter and spring to mature in just under a year. There seems to be a similar breeding pattern throughout the British Isles in spite of the wide range in climate. Hunter and Symonds (unpublished) sampled populations in the east of England and South Wales and found that the breeding seasons occurred at much the same time as in the north of England (Hunter 1968a) even though some of the South Wales slugs in particular were mature up to six months before the main breeding season occurred.

Courtship of mature slugs lasts between thirty minutes and ninety minutes. The spermatophore is variable in shape, but there appear to be two types. It may be a simple chitinous tube of about 0·5 mm long, curved at the slender pointed posterior end and slightly shouldered at the anterior end. Alternatively, it may be disc-shaped at the anterior end with a toothed ridge running along the convex edge of the curved body. The eggs are small (2·5×2mm), white and opaque and are laid in clutches of up to fifty with second and third smaller clutches following at two- to three-week intervals. The eggs hatch in about three weeks in summer and in about two months in winter, depending on the temperature.

This species is found in all damp habitats, in gardens, hedgerows and fields. It is frequently found in large enough numbers to become a pest of agricultural and horticultural crops such as wheat, potatoes and brassicas. High densities (100–200 per sq metre) are not infrequent particularly after the breeding season. The species is however not easy to observe and numbers are often underestimated. *Arion hortensis* can burrow under the soil surface and is frequently found 20–30 cm underground, particularly during dry or frosty weather. When major cracks or spaces occur in the soil these slugs may penetrate to a depth of one metre or more.

There has been considerable confusion over the naming of the slugs *Arion fasciatus* and *A. circumscriptus*. In the present account these slugs will be described as a single species, *Arion fasciatus,* (**Nilsson, 1822**), although there is probably a complex of several species. The length of the extended slug is extremely variable but is usually about 3–5 cm. The colour is also variable; a light silver grey type with dark lateral bands (perhaps to be separated into the distinct species *Arion silvaticus*); a darker grey type, flatter and with less pronounced lateral bands (perhaps to be called *Arion circumscriptus*); and a larger, brownish-yellow type with a yellow zone above the foot *(Arion fasciatus)*. There is so little known about the biology and ecology of this complex that, for most practical purposes, they have to be regarded as one species. The differences between these types will be reviewed by Walden (in press).

The *fasciatus* complex can be distinguished from *Arion hortensis* by the opaque white foot and the bell-shaped contour in transverse section when contracted. The right lateral pigment band arches over the respiratory orifice. A line of larger, paler tubercles along the dorsal side of the body forms an apparent keel which is particularly prominent in the young but which may disappear as the slugs grow older. The life cycle is not known. Breeding probably occurs in the summer months. The eggs, about 3×2mm, are ellipsoidal and hatch in four to five weeks during the summer. This species is common in fields, hedges and woods but appears to be less common in gardens and regularly cultivated land. It is distributed throughout northern Europe (including Iceland) and much of North America.

A second arionid sub-family, the Anadininae, have a penis, either functional or much reduced. The cephalic retractor muscles are separated at their origin and the second posterior loop of the intestine extends beyond the first. There are three genera. *Anadenus* has several large species living on the southern slopes of the Himalayas, extending west to Kashmir and east to China. *Prophysaon* is widely distributed in America. It includes the species *Prophysaon andersoni*, (**Cooper 1872**) (Fig.16), a reddish-grey slug of about

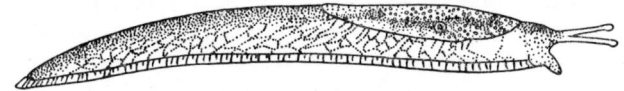

Fig. 16 *Prophysaon andersoni* (after Pilsbury 1948).

6–10 cm which is often a garden pest in California. Other species are found in Washington and Oregon. The genus *Anadenulus* contains only one species, *Anadenulus cockerelli*, (**Hemphill 1890**), restricted to California.

The third sub-family, the Ariolimacinae, is exclusively west American. The penis and epiphallus (an enlarged part of the vas deferens) are well developed and the cephalic retractor muscles are inserted together at the posterior border of the lung floor. The second posterior loop of the intestine extends further backwards than the first. This sub-family includes the genus *Ariolimax*, which are large slugs (18–26 cm when extended) with a conspicuous foot. The genus *Hesperarion* contains smaller slugs (2–5 cm long) which are otherwise similar except that the caudal gland is a deep open pit, not filled with spongy tissue. The genus *Zacoleus* contains small slugs (1·5–2·5 cm) having the breathing pore towards the rear right side of the mantle. There is only one species, *Zacoleus idahoensis*, (**Pilsbry 1903**). *Binneya* and *Hemphillia* are not really slugs as they have an external shell covering part of the mantle and the viscera are confined to the hump, the rear of the foot being solid.

Family PHILOMYCIDAE

These are aulacopod slugs in which the mantle covers the whole back (Fig.17). There is a large shell sac but no shell and the foot sole is undivided. The genital orifice is on the right side of the head. The cephalic retractor muscles are completely separate and are inserted towards the ventral side of the lateral body walls (not dorsally as in Arionidae).

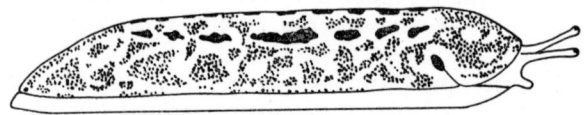

Fig. 17 *Philomycus carolinianus* (after Burch 1962).

These are largely tropical slugs, found from Japan, China and Java to tropical America but some species have become established as far north as Canada. In general they resemble the Arionidae and are probably an earlier branch from the endodontid stock.

Oriental species are separated into the genera *Meghimatium* and *Incilaria* and the American species into *Philomycus* and *Pallifera*.

Philomycus is quite distinct from the other genera as it possesses a special calcified stimulating organ (analogous with the 'dart' apparatus of some snails) situated laterally on the vagina. They are inactive and live in humid woods under the bark of trees feeding on fungi. *Philomycus carolinianus,* **Bosc 1802**, is a buff-brown slug of 5–10cm with dark spots running down the back. It is found from Maine to Florida and west to Iowa and Texas. *Philomycus carolinianus flexuolaris,* **Rafinesque 1820**, with no longitudinal rows of dark spots, is found in Virginia.

Pallifera is similar externally to *Philomycus* except that the head region is somewhat longer and the mantle is free for a short distance at the anterior end. The number of species is unknown but there are several in the eastern and southern states of the USA.

Family LIMACIDAE

Limacid slugs have evolved from one of the most primitive branches of the pulmonata. All of these slugs have a small calcareous shell enclosed by the mantle. There is a distinct keel on the dorsal surface which may be restricted to the posterior end. The foot is usually divided into three separate longitudinal zones. The jaw is smooth, usually with a median projection (oxygnath) and the lateral radular teeth are tri- or bicuspid, the marginal teeth aculeate. The spermathecal duct is short, generally arising from the genital atrium. Two sub-families, the Parmacellinae and the Limacinae, are usually recognised. In the Parmacellinae the shell is either spiral, not completely enclosed by the mantle (in *Parmacella*) or flat and enclosed (in *Milax*). The mantle is granular, with a horseshoe-shaped groove and the respiratory orifice is located behind the middle of the right margin. The genital orifice lies between the respiratory orifice and the right upper tentacle. There is a keel from the rear of the mantle to the posterior end of the slug. The central and lateral radular teeth are tricuspid. The intestine has one forwardly directed loop and there is no rectal caecum. The epiphallus secretes a spermatophore. Glands are present in the wall of the genital atrium. There are three genera, *Parmacella*, *Boettgerilla* and *Milax*, the latter probably being the least primitive. In *Milax* the shell, which is completely enclosed by the mantle, has a non-spiral median nucleus towards the posterior margin. The pedal mucus gland lies free in the body cavity, and is not embedded in the muscle of the foot as in other genera. The median area of the sole of the foot bears inverted V-shaped grooves and there are no lateral body bands. The right ocular retractor muscle lies to the left of the genitalia. There are a large number of species found in many parts of the world, several of these being extremely common. Two species, *Milax budapestensis* and *Milax sowerbyi*, will be described as examples in this account.

Milax budapestensis, (**Hazay 1881**), is probably the commonest species of *Milax* in northern Europe (Fig.14). It is usually slender and, when extended, about 5–10cm in length. The colour is commonly grey to black, the keel being somewhat lighter. The sole of the foot has a central dark zone with lighter areas to each side. The mucus is white. When contracted, the slug curls up into a sickle-shaped curve and is less contracted and humped than other *Milax* species. The life cycle of this slug has been established in various parts of England and Wales (Hunter 1968/9; Symonds, unpublished). Mating takes place in the autumn or spring and is occasionally protracted to twenty-four or even thirty-six hours. During this process curved, toothed spermatophores (about 15mm long) are exchanged. The eggs, laid in batches of up to thirty, are round and opaque, each approximately $3 \cdot 5 \times 2 \cdot 8$ mm.

The speed of development varies with temperature, very little development occurring below 5°C. Since soil temperatures in northern Europe do not rise above this level for considerable periods during the winter, the eggs laid in the autumn do not hatch until the spring (p. 132). The newly hatched slugs are white, and darken to assume their adult colouration in about a month.

Milax budapestensis is common in fields and gardens, often assuming pest status and is distributed throughout northern and eastern Europe. They are good burrowers and can penetrate deep underground. They eat a wide range of growing and decaying vegetable matter.

Milax sowerbyi, (**Férussac 1823**), is large, 7–10 cm when extended, and is brown in colour with darker patches and speckles. The keel is distinctly paler than the rest of the body and is often yellowish-orange. The sole of the foot is not distinctly tripartite and is cream or yellow. The slime is yellow and sticky. When contracted this species does not curl up like *Milax budapestensis,* but contracts into a hemisphere.

The life cycle has not been fully established, but mating and egg-laying are known to occur in the autumn. The large eggs (4×3.5 mm) are often laid early enough to develop and hatch before the onset of winter. The young are creamy white in colour with the mantle a speckled black. They usually grow to maturity in about a year, but this period may be variable.

Milax sowerbyi is found in cultivated gardens and fields but seems to be more restricted in distribution than *Milax budapestensis.* It rarely becomes a pest of agricultural crops.

There are two further complexes of *Milax* species. *Milax gagates,* (**Draparnaud 1801**), is smaller than *Milax budapestensis* or *Milax sowerbyi,* has a colourless slime and lives in gardens and wild places. It is found throughout the temperate regions of the world. *Milax insularis,* (**Lessona and Pollonera 1882**), is larger, blacker than *Milax gagates,* and is common as far north as Britain. Neither of these complexes of species has been adequately defined.

The sub-family Limacinae contains slugs whose keel rarely extends forward as far as the mantle. The nucleus of the enclosed shell is terminal and lies to the left of the mid-line. The mantle is concentrically ridged. There is no well developed epiphallus, no spermatophore and no atrial gland. The stimulatory organ, when present, is a penial sarcobelum, not an atrial stimulator as in *Milax.* The endocones of the lateral radular teeth, when present, are united with the mesocones. The intestine has forwardly directed loops. There are a number of genera, many of them distributed throughout the temperate regions of the world.

The genus *Agriolimax* is known in many parts of the world as *Deroceras,* but this name has been shown to be unacceptable (Watson 1943). *Agriolimax* cannot easily be distinguished from other genera in this sub-family by external features alone. The clearest diagnostic feature is that the rear of the foot is truncated, not sloping gently to a point as in *Limax*.

Agriolimax reticulatus, (**Müller 1774**), is probably the commonest of all species of slug (Figs. 1 and 13). The extended length is about 3–4cm. The colour is extremely variable from a bluish black to the occasional albino, but is usually a general greyish brown to yellow. Darker grooves and patches are scattered over the dorsal surface. The foot is usually cream with a darker zone towards the centre. The mucus is white and sticky. The slug contracts into a hemisphere.

The life cycle of this species has not been fully established in all parts of the world. In northern Europe breeding continues throughout the year but there are usually recognisable peaks of egg-laying (Fig.18). When slugs are confined in cultures in the laboratory or

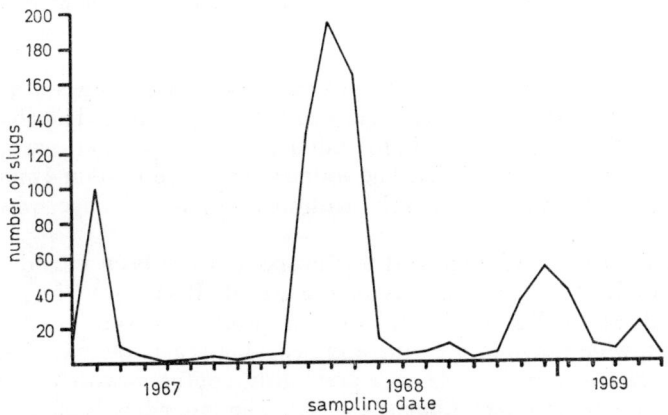

Fig. 18 *Agriolimax reticulatus*. Numbers of newly hatched slugs in field samples (Cambridge 1967–9). Samples were taken every 4 weeks.

outdoors, they attain maturity in a short time (between two and six months). There is, however, some evidence (Hunter and Symonds, unpublished) that growth in the field is much slower and many of these slugs may take up to a year to complete their life cycle, i.e., there may be several overlapping generations in a population at any one time (Fig.19). About 300 eggs are laid by each slug (occasionally more, often fewer, especially in artificially maintained cultures)

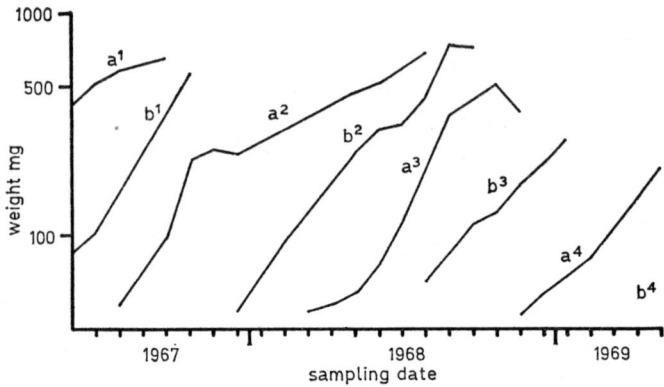

Fig. 19 *Agriolimax reticulatus*. Growth rates of slugs derived from field samples (Cambridge 1967–9). Samples were taken every 4 weeks.

usually over a period of one to three weeks. These eggs are about $3 \times 2 \cdot 5$ mm, translucent with a few calcareous granules. They hatch in about two to three weeks in the summer, but take up to three months to hatch during the winter. The young are pale grey, about 4 mm long. The rate of development varies with weather, taking longer when it is cold or dry.

This species is indigenous to Europe, but has been introduced to most other temperate parts of the world. It is found in gardens, woods and fields and is the most important pest species of slug. It eats a wide range of growing and decaying vegetable material.

Agriolimax reticulatus was previously confused with the species *Agriolimax agrestis,* (**Linnaeus 1758**). The latter has been shown to be a distinct species (Luther 1915). It is slightly smaller, more slender and smoother than *Agriolimax reticulatus* and is pale greyish yellow or nearly white in colour. The distribution of this species is unclear but it is rare in Britain. It is probably widespread in other parts of northern Europe, extending northwards to Iceland. *Agriolimax laevis,* (**Müller 1774**), probably a complex of species, is small, about 2 cm when extended, light or dark brown or occasionally greenish in colour. The mantle is more centrally placed than in other *Agriolimax* and the mucus is clear. It lives in damp habitats and is distributed throughout northern Europe, America, and New Zealand. *Agriolimax caruanae,* **Pollonera 1891**, is about 3–4 cm long, the neck and head being extended somewhat in front of the mantle. The colour is a translucent chestnut brown, flecked with darker patches. The foot is grey, the mucus is thin and colourless. Probably the most obvious difference between this species and *Agriolimax reticulatus* lies in its much greater activity and faster movement. This species occurs throughout northern Europe, mainly in the western part, and it has also been introduced into north America (California).

In slugs of the genus *Limax* the rear of the body slopes gently and evenly to a point, it is not truncated as in *Agriolimax*. The nucleus of the concentric mantle ridges lies in the mid dorsal line, not to the right as in *Agriolimax*. *Limax maximus*, **Linnaeus 1758**, is a large slug, 10 to 20cm when extended (Fig.15). It is brown or grey with large patches or spots of darker pigment on the mantle. The tentacles are translucent, pinkish brown. The body has two or three darker bands running from the back of the mantle to the tail. The foot is uniformly pale. The mating procedure of this species is complex (p. 89). *Limax maximus* occurs in woods, hedgerows, gardens or outhouses. It seems to feed mainly on fungi and decaying matter. It is not often common but is found throughout Europe, Asia Minor, Algeria and the Atlantic Islands. It has been introduced into North America, South Africa, Australia and New Zealand. *Limax flavus*, **Linnaeus 1758**, is a greenish-yellow colour with yellow mottles on the mantle. There are no lateral bands and the tentacles are a very characteristic steel blue. These slugs are frequently found in cellars and drains, in broken walls and outhouses. Occasionally, they crawl up drains, into bathrooms and kitchens. They also occur in woods and gardens. The species is distributed throughout Europe and eastwards to Syria and it has been introduced into South Africa, Australia, North and South America and to the Atlantic and Pacific Islands. *Limax cinereoniger*, **Wolf 1803**, is the largest slug species, growing to 20 or 30cm in length. The keel is prominent and long and the slug is usually entirely black, except for a lighter mid-dorsal line and central zone of the foot. Single individuals have been known to survive for over five years (Oldham 1942). This species inhabits woodlands and hill sides and is rarely found in cultivated areas. It is distributed widely throughout Europe.

Slugs of the genus *Lehmannia* tend to have a more translucent appearance than *Limax* and there is a long rectal caecum. They readily absorb water and become quite swollen when kept in a moist environment. *Lehmannia marginata*, **(Müller 1774)**, is common in woodlands, rocks and stone walls, preferring damp habitats. It is commonly noted climbing trees, and feeds on fungi and lichens—hence its common name, the 'tree slug'. It occurs in northern Europe, particularly to the west, but has been introduced into the east coast of America. *Lehmannia poirieri,* (**Mabille 1883**), probably originated in Spain but has been introduced into California and is found in glasshouses in northern Europe. It is paler and more yellow than *Lehmannia marginata* with the body bands closer to the mid-dorsal line.

Family ATHORACOPHORIDAE

The Athoracophoridae (Fig. 20) have only one pair of tentacles, and

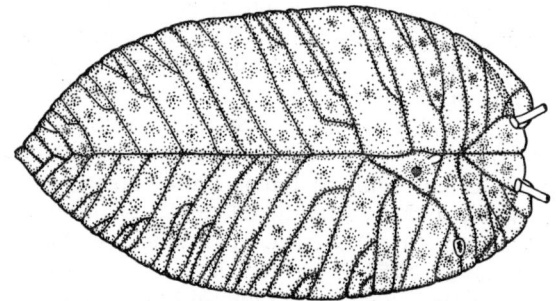

Fig. 20 *Pseudaneita papilata* (after Burton 1962).

the small mantle area contains an internal rudimentary shell of calcareous deposits. There is a distinct pattern of dorsal grooves and a dorsal tracheate lung of a nature found in no other mollusc. This lung consists of a pulmonary cavity, with thin-walled pulmonary diverticula arising from the base and radiating to form the roof of a dorsal blood sinus. The slugs grow up to 15 cm in length and are often more attractively coloured than other slugs. They frequently have dorsal papillae. The anatomy of these animals is so different from that of the other heterurethran snails that there are strong arguments for creating a separate order for the Athoracophoridae (Burch 1968).

These slugs are common in both bush and grassland in New Zealand, New Caledonia, New Hebrides, New Britain, The Admiralty Islands and the eastern coast of Australia. They are frequently found under logs, leaf mould and in the leaf bases of flax bushes and nikau palms. The slugs eat only fungi. They breath mainly through the skin of the back which is kept moist by a renal secretion passing along the grooves. The role of the lung in respiration is not yet known.

The eggs are round, gelatinous, papillate and a light translucent yellow. They vary in diameter from 3 to 7 mm according to species. They are laid in batches of up to fifty in damp, cool surroundings from the beginning of spring until the late summer, and take up to three months to hatch.

There are two sub-families. The Athoracophorinae, with the foot separated from the body at the lateral edges by an hyponotum were, until recently, lumped together in the genus *Athoracophorus*. However, Burton (1963b) has separated these slugs into four genera, largely on the size and shape of the mantle area. The Aneiteinae, with three genera, have no hyponotum.

Family VERONICELLIDAE

These slugs (Fig. 4) have two pairs of tentacles, but the pair possessing the eyes is contractile (Fig. 5a) not inversible as in most slugs and the other tentacles are bilobed. They have a large mantle, which contains no shell, covering the entire back. The foot is transversely grooved and is separated from the rest of the body by a groove. The sides of the mantle (hyponota) are separated from the rounded back by a sharp ridge, the perinotum.

The anus and pneumostone open below the hyponotum behind the foot. The male genital opening is on the right side of the body, in the groove separating the foot from the mouth, and the female opening is towards the middle of the hyponotum. The radular teeth are unicuspid, decreasing in size from the centre towards the edges. There is a curved jaw made up of numerous, small parallel plates. The nervous system is concentrated.

These slugs are nocturnal herbivores. They can be pests in the tropical regions of America, Africa and Asia, to which their distribution is limited. There are three main genera.

The biology of the Veronicellidae is not well known. The Argentinean species *Vaginulus borellianus,* (**Colosi 1921**), however, has received some attention (Lanza and Quattrini 1964). There appears to be a period of activity between October and April/May with no reproduction in other months. When kept in a constantly humid atmosphere at 18–30°C they can reproduce throughout the year. They live for about a year to eighteen months, attaining a weight of 15–23 g by ten to eleven months. They can mature at about three to seven months laying their eggs (about 500–1,500 in eight to thirteen batches of

25–200 eggs) in a mucus envelope on the soil surface. There is some dispute about the frequency of cross-fertilisation. Lanza and Quattrini believe that copulation is rare. Most species are found in the South Pacific region, but the large, oblong grey slug, *Veronicella floridana,* **(Leidy 1851)**, occurs in the United States.

Family RATHOUSIIDAE

These slugs are very similar to the Veronicellidae except that the anus and renal openings are on the right side above the female opening, and there is no jaw. Very little is known about these animals. They feed on plant and animal material. The two genera *Rathousia* and *Atopos* are found in South-East Asia and Australia.

3

FOOD, FEEDING AND DIGESTION

FOOD

Slugs feed on leaves, stems, bulbs, and tubers, but they also consume fungi, lichens, algae and animal material. Many anecdotal observations have been made on what slugs will eat both in the wild and in the laboratory, e.g., pigeon's eye-balls and snakes; but these would not be expected to form a large part of the diet of *Agriolimax reticulatus*. Very little systematic information is available on the make-up of the diet of populations in the wild, even less on what nutrients are necessary and in what quantities for normal growth.

One method of finding out what slugs eat is to analyse the fragments of plant material in the digestive tract and faeces. The major dietary constituents in *Agriolimax reticulatus* collected from a small oakwood (Pallant 1969) were the green leaves of the creeping buttercup and the stinging nettle (Table 2).

There were seasonal variations in the relative amounts of these two main food plants. When given a choice of diet in the laboratory the preferences were similar but animals that had been feeding mainly on stinging nettle in the wild sometimes preferred buttercup leaves and vice versa. This perhaps indicates that the type of food found in the digestive tract reflects the relative abundance of that plant in the immediate vicinity of the slug as well as any food preference. These studies were carried out on mature slugs only and no chemical analyses of the foods have been made.

Another way of studying slug nutrition is by using artificial diets. Frömming (1957) observed that *Limax flavus* was often found as a pest of stored foods in the cellars of old houses and in granaries. In the laboratory, this slug completed its life history on a diet of flour, but its growth rate and the size attained were less than normal. When *Arion rufus* was reared on this diet not only did it complete its

TABLE 2
THE CONSTITUENTS OF THE GUT CONTENTS OF *AGRIOLIMAX RETICULATUS* (PALLANT 1969)

Constituent	% of slugs
Creeping buttercup *(Ranunculus repens)*	53·4
Stinging nettle *(Urtica dioca)*	23·9
Ground ivy *(Glechoma hederacea)*	2·0
Wood speedwell *(Verbena montana)*	1·3
Wood sorrell *(Oxalis acetosella)*	0·6
Unidentified dicotyledon	21·2
Unidentified monocotyledon (mainly grass)	12·3
Moss leaves	3·4
Worm chaetae	1·3
Arthropods (mainly aphids and small diptera)	21·2

life history but its growth rate and size at maturity were also normal; it even lived longer. In both species the diet of flour accentuated the red pigmentation in the body wall. A more synthetic diet has been studied by Ridgway and Walker (personal communication), based on one developed for insects and containing protein, carbohydrate, fats, vitamins, salts and an antibiotic. *Arion ater* grew and developed normally on this diet. So far these authors have studied only the effect of omitting vitamins of the B group from this diet. All members of this group of vitamins appear to be essential for growth, for lack of any one of them resulted in a gradual falling off in the growth rate, frequent cannibalism, and marked reduction in the length of life.

FEEDING

Like most other gastropods, slugs feed by means of a radula. The action of the radula can be studied by several methods.

Food, feeding and digestion

The older naturalists noted than when slugs had been feeding for example on algal films, or lime washes on green-house windows, a clear feeding track resulted, because of the removal of food material by the action of the radula. German workers, most recently Märkel (1957), have studied feeding experimentally by providing artificial food films on glass. These substrates are made by coating glass with a gelatin suspension of finely ground maize meal or fat, and letting it dry. Slugs placed on this film consume it avidly and when the plates are later stained with iodine (staining the starch in the remaining maize meal a dark blue) or oil red 0 (which stains fat red) the marks made by the radula can be studied in detail. An even simpler technique is to allow slugs to feed on the emulsion layer from well washed photographic negatives. In favourable conditions, the marks of the individual teeth are visible (Fig. 24). While this method can yield valuable results, thin films (e.g., encrusting algae) are not part of the natural diet of most slugs. At present feeding on a natural food can only be studied by killing the animal a short time after feeding and examining the recently ingested food. The size and shape of the food particles in the gut gives some idea of the way in which the radula is used. Both of these methods have to be supplemented with a study of the morphology of the radula and associated organs. Perhaps only by using all these techniques, together with microcinephotography, will we come fully to understand feeding in these animals.

The buccal mass is best understood by reference to a sagittal section (Fig. 21) through the head. A cuticle-covered epithelium lines the lips and buccal cavity. The jaw and radula can be regarded as specializations of this cuticle. The downward projecting ventral edge of this jaw is usually dark in colour and hard. Along the roof of the buccal cavity there is a groove, termed the dorsal food groove which leads into the oesophagus, and along the lateral walls of this groove there are two longitudinal bands of ciliated cells. The large odontophore fills most of the buccal cavity. The anterior and dorsal surface of the odontophore is covered by the radula, the remainder being covered by the cuticularised epithelium. While the odontophore cartilage of prosobranch gastropods histologically resembles vertebrate cartilage, that of pulmonates has become considerably modified. It consists of many muscle cells interspersed with large, thin-walled vacuolated cells. The cartilage is in the form of a U-shaped gutter with the anterior end considerably tapered. For most of its length the radula, similarly U-shaped in section, lies in the dorsal groove of the cartilage (Fig. 8a), but at its anterior end it is reflected over the tapered end of the cartilage and is continuous with the buccal cuticle. The posterior two thirds of the radula are contained within a radular gland and the cavity of the U is filled with a rod of peculiar connective tissue termed the collostyle. On dissection, the rounded tip of the radular gland can be seen protruding from the posterior end of

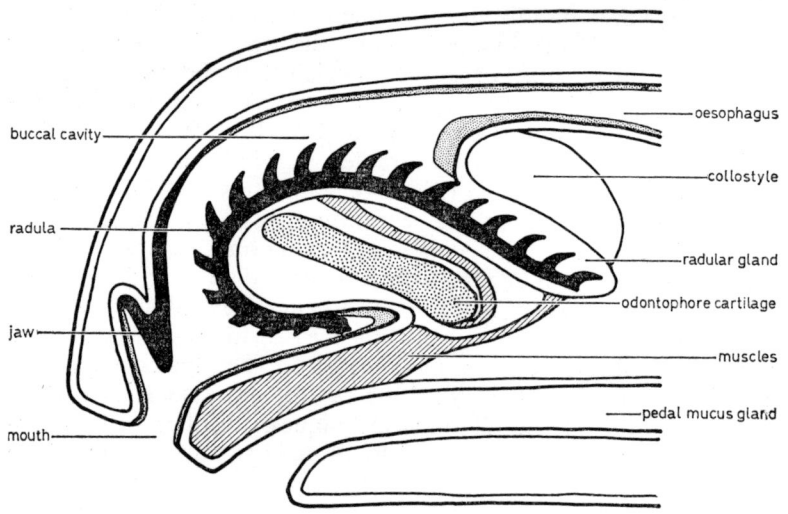

Fig. 21 Sagittal section through the head of a slug.

the buccal mass. The odontophore contains a large number of muscles and as Carriker (1946) stated in an extremely thorough account of the anatomy of the pond snail *Lymnaea stagnalis appressa,* 'the musculature of this specialized organ is the most complicated in the entire animal and indeed is one of the most intricate muscular structures in the invertebrate group'. Carriker describes twenty-eight different muscles, many of them paired, within or connected with the buccal mass. The musculature of the buccal mass of *Arion ater* appears to closely resemble that of *Lymnaea,* (Roach 1966).

The radula is usually studied by dissecting out the buccal mass, and boiling it in 10% potassium hydroxide. This destroys the soft tissues and leaves behind the chitinous radula. After washing the radula is mounted in polyvinyl lactophenol containing the stain lignin pink. The outlines of the teeth are clearly seen under the light

Fig. 22 *Agriolimax reticulatus* radula in feeding position. Scanning electron microscope. The lower pictures show details.

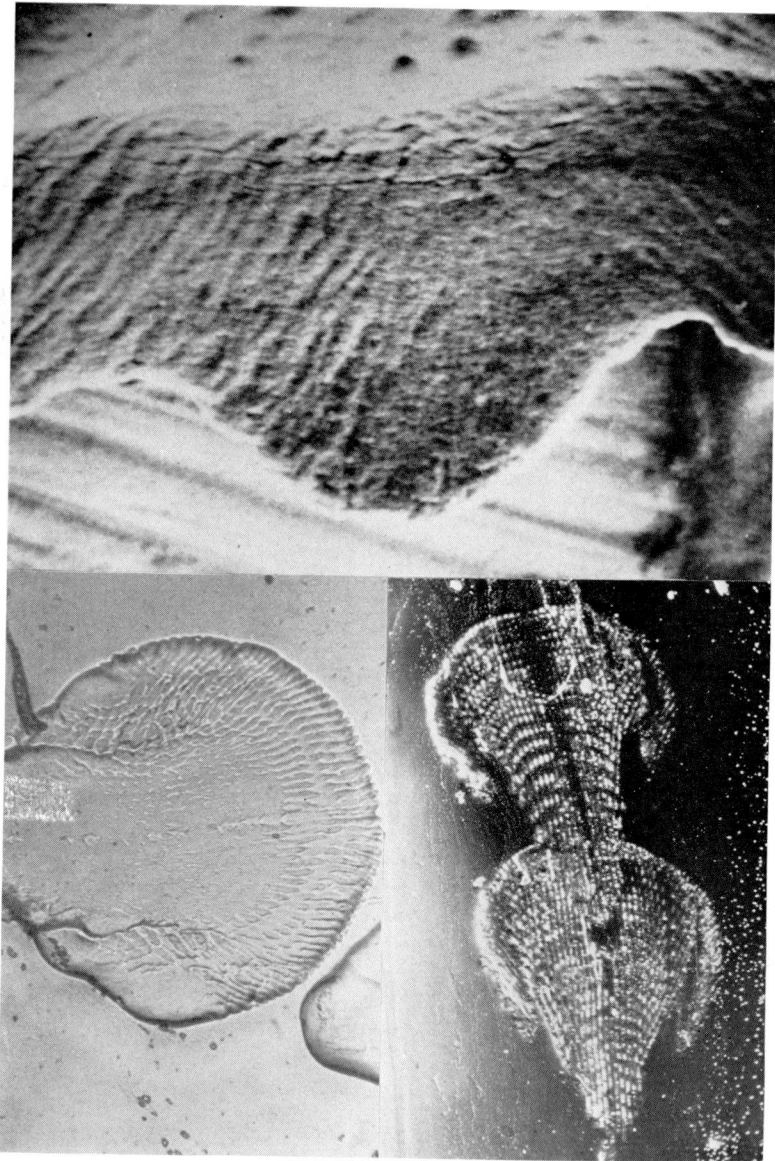

Fig. 23 (ABOVE) The edge of the jaw of *Agriolimax reticulatus* showing the marks caused by the radula teeth. Scanning electron microscope.

Fig. 24a and b (BELOW) *Agriolimax reticulatus*. Feeding tracks in carrot juice agar. The marks of the individual radula teeth and their direction of movement are easily visible.

Food, feeding and digestion

microscope but the appearance is essentially two-dimensional. Teeth are arranged on the radula in the form of regular longitudinal and transverse rows. Each transverse row has a central tooth, flanked on each side by about twenty lateral teeth, and these in turn are flanked by about twenty marginal teeth. The number and size of the teeth increase with the age and size of the animal. Across each transverse row there is a gradual change of shape of the teeth but the transverse rows closely resemble each other so that the teeth in each longitudinal row are almost identical. Occasionally a radula is found where one or more teeth are deformed; then usually all the teeth in that longitudinal row have an identical abnormality.

More recently radulas have been examined in the scanning electron microscope (Runham and Thornton 1967; Runham 1969). This has a considerable depth of focus and a much higher resolution than the light microscope so that it is possible to study in detail the three-dimensional structure of the teeth (Fig. 22).

The sequence of events during feeding has not been completely established but appears to be as follows. The surface of the potential food is examined by the anterior tentacles and mouth lobes. If acceptable, the mouth is applied to the surface of the food and is opened by the expansion of the lips. The radula-covered odontophore is then swung forwards and downwards to meet the substrate. A forward movement of the odontophore tip constitutes the feeding stroke. At the end of this stroke the odontophore is withdrawn into the buccal cavity. The jaw also aids feeding. When actively feeding on a flat surface the animal executes a number of these feeding strokes while moving the head from side to side and slowly moving forwards.

During the feeding stroke the radula is moved over the cartilage. As the radula is retracted over the edge of the cartilage the teeth change their direction of movement. On the outside of the odontophore the teeth move anteriorly and approximately horizontally but as they move over the edge the teeth swing through 240° to enter the groove. The teeth cusps are erect on the slightly convex exterior of the odontophore but in the groove they lie flat (Fig. 22). It is usually stated that the radula is used like a rasp and this must indeed be the main action of the teeth on the outside of the odontophore, but examination of the crop immediately following a meal reveals that large chunks of material are present in addition to the expected small particles. The way in which these large particles are obtained is at present uncertain but several methods are possible. As the teeth pass over the edge of the cartilage, each row functions like a chain-saw and the inward swing of the cusps at this point may serve to tear off chunks of food. That the jaw bears clear evidence of radula action (Fig. 23) would imply that it has some important function during feeding. It could either serve as a hard surface against which the radula cuts, or, alternatively, its hardened edge might act as a cutting edge.

Testacella on contact with a worm pushes the odontophore out through the mouth (Webb 1893). The walls of the groove in the odontophore, which are covered with backward pointing needle-like radular teeth (Fig. 25), diverge and clamp round the body of the

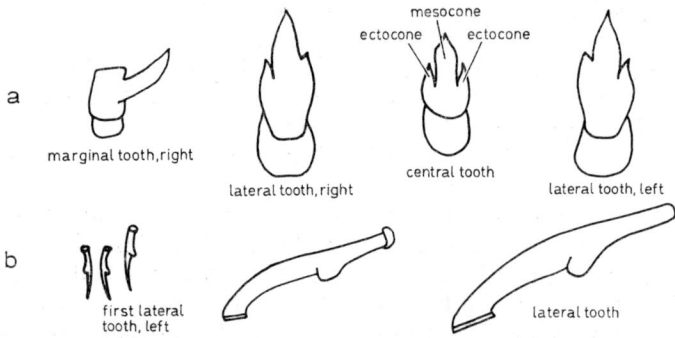

Fig. 25 Radular teeth.
    a. *Agriolimax reticulatus*.
    b. *Testacella scutulum*, lateral teeth, no central tooth is present in this animal (after Taylor 1907).

worm. Webb likened this action to that of the steel jaws of a spring trap. The worm is dragged in through the mouth in stages and slowly ingested; this may take several hours for a large worm.

It is uncertain how the food passes from the buccal mass into the oesophagus. Large pieces of food could be transported by peristalsis in the buccal mass but the smaller pieces have first to be removed from the spaces between the radular teeth. It is possible that the dorsal ciliated bands of the food groove and perhaps also the saliva are important here.

Feeding results in the wearing down of the teeth cusps, and examination of the anterior edge of the radula shows that teeth become reduced to small rounded stumps. Patches of the radula become detached at this edge and are swallowed. The rate of loss of teeth can

Food, feeding and digestion

be studied by collecting the faeces and counting any teeth in them. More accurate estimates can be obtained by various experimental methods which were devised after a detailed study of the radular gland (Runham 1962, 1963 and Isarankura and Runham 1968). At the hind end of the radular gland there is a group of cells which secrete the radula—the odontoblasts. The radula is formed by cementing together very long cellular processes, or microvilli, from these odontoblasts with secretions from the odontoblasts and surrounding epithelial cells. Immediately around the odontoblasts there is an area of very actively dividing cells which gives rise to the epithelia over (superior epithelium) and under (inferior epithelium) the newly secreted radula. When first secreted the radula is soft, but as it passes forward secretions from the superior epithelium harden the teeth while the secretions from the interior epithelium make the membrane tough but flexible. The rate of secretion of new teeth was first studied by destroying small areas of the odontoblasts and following the forward movement of the area lacking teeth by examining the radulas of animals killed at intervals after the operation. Injections of colchicine to inhibit cell divisions also gave rise to abnormal rows of teeth as did the anaesthetic used for the operations (magnesium chloride or sulphate). In *Limax flavus* X-rays (8,050–130,000 Röntgen) were found to damage the dividing cells in the radular gland and produce changes in the odontoblasts, leading to the production of abnormal rows of teeth (Kerth and Krause 1969). Collections of animals from the field following the severe winter frosts of 1963 indicated that cold shocks could give rise to the production of abnormal rows of teeth and from this observation a versatile method for following radula production was developed. When animals are placed at 0–1°C for one to two days and then brought out into normal temperatures it is found that there are one to five rows of abnormal teeth in 95% of the animals. This method was successful with all pulmonates tested no matter what their size. All the methods give comparable results and the rate of radula production is found to vary from species to species and to depend on age, size, and temperature (Table 3).

The abnormal teeth obtained with the use of such methods were found to move forwards until eventually they separated from the anterior end of the radula. Radioactive labelling, using tritiated thymidine, showed that the epithelial cells moved forward with the radula. From a close study of the cytology of the normal radula, the effects of the operations, and from transplantations of the whole radular gland, it was shown that several processes are involved in

TABLE 3

Species	Temperature °C	No. of rows of teeth produced per day
Arion ater (mature)[1]	20	3·9
Agriolimax reticulatus (mature)[1]	20	5·1
Agriolimax caruanae (mature)[1]	20	5·6
Limax flavus (48 days old)[2]	15	3·1
Limax flavus (1–2·5 yr old)[2]	15	1·4

[1] Isarankura and Runham 1968
[2] Kerth and Krause 1969

radula production and loss: (1) secretion of new radula; (2) movement forward of the radula and epithelia; (3) hardening of the radula and breakdown of the upper epithelium; (4) separation of the worn radula from the underlying epithelium and (5) detachment of chunks of the separated radula during feeding. Normally all of these processes are closely integrated so that there is a steady increase in size of the radula with age and a fairly constant size of radula for a given size of animal. This integration can in exceptional circumstances be upset. Thus when *Helix aspersa* were kept for long periods (three months) in a dormant state, there was an abnormal lengthening of the radular gland although the rate of production of new teeth was normal (Isarankura and Runham 1968).

DIGESTION

After surveying the digestive systems of a very large number of different animals Yonge (1937) concluded that the gut could be subdivided into a number of functional regions—reception, conduction, storage, digestion, absorption, faeces formation and faeces conduction. All of these functions are carried out by the slug digestive system, but several functions may proceed at the same time in one morphological region.

The digestive system can be studied using a wide variety of techniques. As a basis for any study the anatomy of the organs and

Food, feeding and digestion

their histology and cytology must first be determined. The functions of any ciliated areas can be studied with suspensions of particles, e.g., indian ink, carmine, colloidal graphite (Aquadag). If any secretory granules are observed in the cells they can be identified, at least partly, by histochemistry. The cells can also be studied during feeding

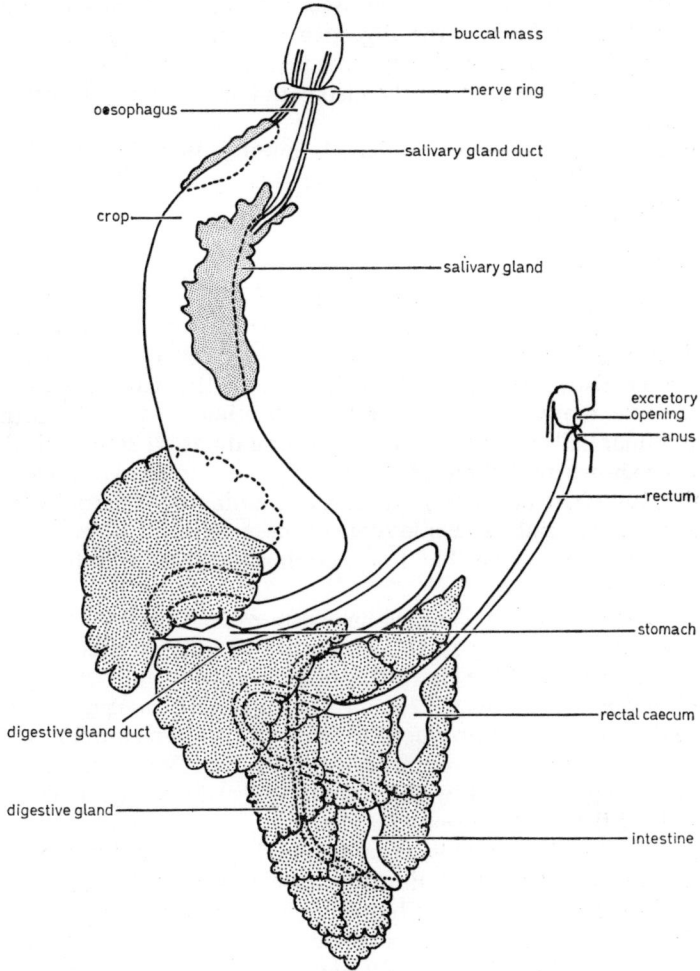

Fig. 26 *Agriolimax reticulatus* digestive system (after Walker 1969).

and starvation to see if any changes in them can be correlated with their activities. The biochemistry of any secretions and of tissue extracts can be determined and the morphological and biochemical changes in any food material can be followed during its passage through the gut. Now radioisotope-labelled biochemicals are so readily available, the fate of ingested food materials can be studied in great detail using extracts of the tissues, or histologically by autoradiography. Studies on the digestive systems of slugs have unfortunately in the main been concerned with one or a few aspects of anatomy, physiology, or biochemistry and no properly integrated study has been made.

The account given here is based, unless otherwise stated, on Walker's (1969) account of the digestive system of *Agriolimax reticulatus* (Fig. 26).

Salivary glands

The salivary glands consist of many sacs of cells (acini) whose small collecting ducts merge to form the large paired salivary ducts. Within the acini there are ten quite distinct cell types which are apparently distributed at random throughout the acinus, with one type a little more common near the collecting duct, and another (the granular cell) localised in the collecting ducts. Histochemically, it can be shown that three of the cell types are mucocytes. The remaining cell types produce secretions containing both carbohydrate and protein materials, so while some of these may also be mucocytes others must be responsible for secreting the enzymes found in extracts of the glands.

As no experimental technique has been evolved for collecting the saliva we know nothing of the properties or volume of the saliva produced. Extracts of the glands have been shown to contain amylase in both *Arion ater* (Evans and Jones 1962a), and *Agriolimax reticulatus*. Where the enzymes are produced in the gland is unknown, but in the snail *Lymnaea stagnalis* Boer, Bonga and Rooyen (1967) have shown that amylase is produced by the granular cells which are localised in the walls of the small collecting ducts.

The secretion is probably moved into the salivary ducts by the cilia present on the cells lining the collecting ducts. No cilia are present in the salivary ducts; but there is a well-developed layer of muscle in the wall. Regular peristaltic waves can be seen in the salivary ducts of *Agriolimax reticulatus,* which probably serve to move the saliva along the ducts into the buccal cavity.

Food, feeding and digestion

The oesophagus

The lining of the oesophagus is thrown into six longitudinal folds consisting of a columnar epithelium, most of which is covered with a thin cuticle. The most dorsal groove is continuous with the dorsal food groove of the buccal cavity roof. Small ridges of ciliated epithelium are present at the anterior and posterior ends of the oesophagus, and the cilia create a current which conveys particulate material up their sides and back along the tops of the ridges into the crop. Mucus cells also occur in the ciliated areas and probably aid in particle transport or in lubrication. Underlying the epithelium there are two layers of muscle, which in *Arion ater* (Roach 1968) give rise to waves of peristalsis.

The crop

This is one of the first organs seen when a slug is dissected. It is a large fluid-filled sac, often yellow, red or brown in colour. The absence of ridges clearly demarcates this region from the oesophagus although it originated as a specialised area of the oesophagus. A columnar epithelium of cells containing granules, apparently secretory, lines the crop. Amongst these cells are patches containing both ciliated and mucus cells, and underlying the epithelium there are two thin layers of muscle. Roach (1968) has shown in *Arion ater* that there are clear waves of peristalsis having a frequency of 1 per 1·2min at the anterior end, and 1 per 3min at the posterior end of the crop. He suggests that the contractions at the anterior end aid circulation of the crop contents, while at the posterior end their function is trituration (grinding). In *Agriolimax reticulatus* there is no evidence of trituration but there is a thickening of the muscle layer at the posterior end which, as X-rays show (see below), functions as a sphincter, (Walker 1969).

The fluid in the crop, called the crop juice, as in other pulmonates, contains a number of enzymes active against a wide range of substrates (Table 4). In *Agriolimax reticulatus* the pH of this fluid was 5·8 when food material was present and pH 7·7 when it was absent. From Table 4 it can be seen that the carbohydrases have been the most extensively studied, but as Evans and Jones (1962a) point out, some carbohydrases are not very substrate-specific so that the number of enzymes may be much smaller than indicated by Table 4. Large numbers of bacteria are present in the crop juice and in *Helix pomatia* some species actively digest chitin and cellulose (Jeuniaux 1954, 1963). It now seems likely that pulmonates are also

TABLE 4
ENZYMES OF THE CROP JUICE OF SLUGS

Enzymes	Substrates tested	Species	Author
Amylase	Starch, amylose, amylopectin, glycogen	*Arion ater, Agriolimax reticulatus*	Evans and Jones (1962a), Stone and Morton (1958), Walker (1969)
Cellulase	Degraded cellulose, cellulose derivatives	*Arion ater, A. fasciatus, Agriolimax reticulatus, Limax flavus*	Stone and Morton (1958), Evans and Jones (1962a), Nielsen (1962)
α-glucosidase (invertase)	Sucrose, maltose, melezitose, methylgalactoside	*Arion ater, Agriolimax reticulatus*	Stone and Morton (1958), Evans and Jones (1962a), Walker (1969)
β-glucosidase	Salicin, amygdalin, cellobiose, gentiobiose p-nitrophenyl-β-glucoside	*Arion ater, Agriolimax reticulatus*	Evans and Jones (1962a), Stone and Morton (1958)
Trehalase	Trehalose	*Arion ater*	Evans and Jones (1962a)
α-galactosidase	Melitose, methyl glucoside, raffinose	*Arion ater*	Evans and Jones (1962a)
β-galactosidase	Lactose	*Arion ater*	Evans and Jones (1962a)
Xylanase	Xylan	*Arion fasciatus, A. ater, Agriolimax reticulatus*	Nielsen (1962), Stone and Morton (1958)

Enzymes	Substrates tested	Species	Author
Laminarinase	Laminarin	*Arion fasciatus, A. ater, Agriolimax agrestis, Limax cinereoniger*	Nielsen (1963), Stone and Morton (1958)
Alginase	Alginic acid, sodium alginate	*Arion ater*	Franssen and Jeuniaux (1965)
Chitinase	Degraded chitin	*Arion fasciatus, A. ater, Agriolimax agrestis*	Jeuniaux (1954)
Pectinase	Pectin	*Limax cinereoniger, Arion fasciatus*	Evans and Jones (1962a)
Cathepsin	Casein	*Arion ater*	Evans and Jones (1962b)
pH7·6 Protease	Peptone, casein, fibrin	*Limax, Arion*	Graetze (1929)
Gelatinase	Gelatin	*Agriolimax reticulatus*	Walker (1969)
Lipase	Olive oil	*Agriolimax reticulatus*	Walker (1969)

able to produce chitinase and cellulase enzymes (Strasdine and Whitaker 1963), but it is still possible that bacteria contribute to the enzyme complement of the crop juice.

Extracts of the crop wall can be shown to contain weak amylase, invertase, lipase and protease activities but it is not clear what contribution these enzymes make to the crop juice. Histochemically it is found that the apical granules in the crop epithelial cells are intensely positive for lipase.

The stomach

This is a small sac-like structure which has a series of folds on its dorsal wall (Figs. 27 and 28). Starting at each of the two digestive gland openings there is a large fold or typhlosole and these run back to converge, forming the walls of the intestinal groove. An anterior triangular accessory fold is present between the converging typhlosoles; the grooves between the typhlosoles and the accessory folds are the two gastric grooves. Radiating out from the digestive gland openings there is a series of small folds which extend out over the stomach wall for a short distance. The epithelium of the stomach wall is columnar and richly ciliated over the folds, the grooves and the posterior wall. When the stomach is opened and particulate material added, the material is drawn towards the digestive gland openings and then transported away from the opening by the cilia on the folds. None of the particles tested was observed to pass into the hepatic ducts. Strings of material, termed liver strings, were observed to emerge from the digestive gland openings into the gastric grooves and were transported back into the intestinal groove, together with other particulate material collected from the stomach lumen by the typhlosoles. The cilia in the posterior part of the intestinal groove, in contrast to those in the gastric grooves and anterior part of the intestinal groove, transport particles anteriorly. Particles accumulate where the opposing ciliary currents meet and this represents the area of faeces formation.

Muscle layers are present in the stomach wall. In *Arion ater* the frequency of the stomach contractions is 1 per 1·5 min (Roach 1968). There are apparently no sphincters at the entrances to the digestive glands or the intestine.

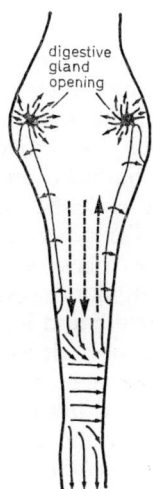

Fig. 27 *Agriolimax reticulatus* stomach (after Walker 1969). The stomach has been opened dorsally to show the ciliary currents (arrows) on the ventral floor.

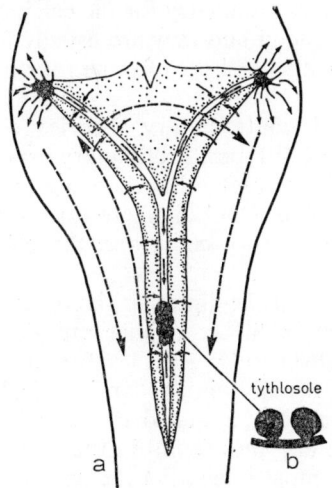

Fig. 28 *Agriolimax reticulatus* stomach (after Walker 1969).
a. The stomach has been opened ventrally to show the ciliary currents (arrows) on its dorsal surface.
b. A section across the typhlosoles.

The digestive gland

The paired digestive glands are immediately obvious on dissection as they occupy a large part of the body cavity and are highly pigmented. One gland is situated anteriorly, the other posteriorly, and they open separately into the stomach by way of hepatic ducts. Each digestive gland is subdivided into lobes, the number and arrangement of which appears to be constant, and in *Agriolimax reticulatus* there are six in the posterior and nine in the anterior digestive glands. Within the lobes there are many acini and the acinar ducts merge to form lobular ducts which then join to form the hepatic ducts. When the hepatic and lobular ducts are opened it is found that there are ciliated folds along their length some of which join with those around the stomach openings. The ciliary currents in the opened ducts are all directed towards the stomach.

Muscle fibres are found around the acini and ducts. In both *Arion ater* and *Agriolimax reticulatus* peristaltic contractions of the lobules and ducts have been observed (Roach 1966, and Walker 1969).

The cells lining the acini are of four types: digestive, calcium, excretory and thin. In *Agriolimax reticulatus* these cells are randomly distributed but there is a tendency for the calcium cells to be located in the corners of the acini and they are usually flanked by thin cells. The cytology and histochemistry of these four cell types is summarised in Fig. 29.

There are several contradictory theories on the relationships between the various cell types. Preliminary experimental work with *Agriolimax reticulatus* suggests that there are two cell lines: one, the digestive cell, which can either divide to form further digestive cells, or transform into excretory cells; and the calcium cells, perhaps arising from the thin cells.

Analysis of extracts of the digestive gland has shown that many of the enzymes present in the crop juice are also to be found here, (Evans and Jones, 1962a and Walker 1969), and that the pH optima of these enzymes are usually similar or identical. Because of these similarities, it is generally assumed that the digestive gland is the major source of the enzymes found in the crop juice. This explanation is certainly the most likely, but the passage of a fluid secretion from the digestive gland through the stomach and into the crop has yet to be discovered. Some of the enzymes in the crop juice could originate from the salivary glands, crop walls, or bacteria. It is also likely that some of the enzymes present in the digestive gland extracts

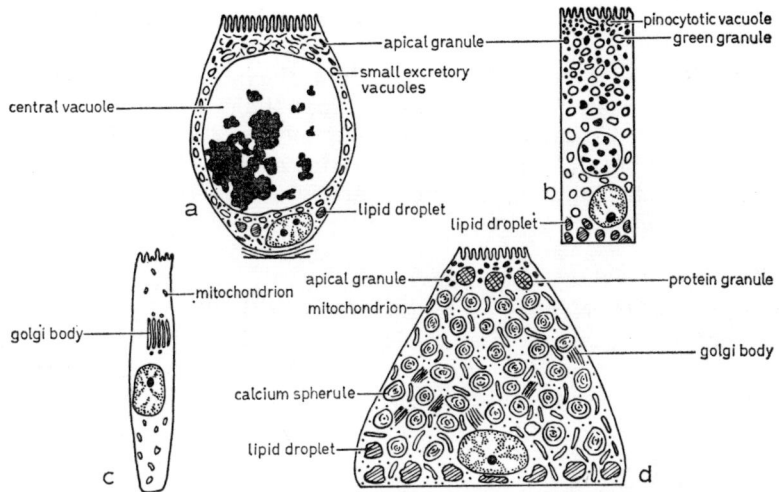

Fig. 29 *Agriolimax reticulatus* digestive gland cells (after Walker 1969).
a. Excretory cell.
b. Digestive cell.
c. Thin cell.
d. Calcium cell.

are in fact intracellular enzymes concerned with intracellular digestion. The digestive gland is usually also held to be the main reserve food storage area. These various functions of the digestive gland will be discussed later.

The intestine

The intestine is coiled between the lobes of the digestive gland, and its arrangement appears to be constant and characteristic of the species. It can be subdivided into three regions, pro-, mid- and postintestine. The prointestine has six longitudinal ridges and is lined by a ciliated columnar epithelium with scattered mucocytes. At the junction with the midintestine the ridges are lost and a characteristic type of cell, the intestinal cell, appears in the epithelium. These cells contain large protein granules of completely unknown function, but it seems possible that they are concerned with the secretion of the sheath surrounding the faecal string. The postintestine lacks the intestinal cells but many mucocytes are present and the wall is thrown into a number of small folds each with a ciliated tract along the summit.

Underlying the epithelium throughout the intestine there are two muscle layers. In *Arion ater*, (Roach 1968), the muscular movements of the isolated intestine are complex but only weak peristaltic contractions have been observed in the intestine of *Agriolimax reticulatus*. The pH of the intestinal contents is approximately pH7 in *Arion ater* and *Agriolimax reticulatus*.

The rectum and rectal caecum

The wall of the rectum has several longitudinal folds; these are small except for one large fold on the lateral wall. Cilia abound and there are also many interspersed mucous cells. A very sparse layer of muscle underlies the epithelium. The histology of the rectal caecum is similar to that of the rectum but its epithelium lacks cilia.

The functioning of the digestive system

Slugs normally become active only towards nightfall, and usually retire again about dawn. For part of the time that they are active they are feeding (Newell 1968). There is thus a diurnal rhythm, consisting of a rest period during the day, and a period of feeding at night.

The time that food remains in the gut has been studied in two ways. Roach (1966) observed that the colour of the faeces was dependent on the colour of the food being consumed. He therefore maintained some *Arion ater* on food of one colour (e.g., carrot) and then gave them a meal of a different coloured food (e.g., cabbage or lettuce). The differently coloured faeces first appeared 12 hrs. 35 min ± 2 hr 15 min later, but the new food was still present in the faeces 25 hr after the meal. Walker (1969) obtained more detailed information by feeding *Agriolimax reticulatus* radio-opaque material and X-raying the animals at intervals thereafter. He used two radio-opaque materials, barium sulphate with a particle size of 2–3 μ, and colloidal thorium dioxide (Thorotrast) with a particle size of 0·1–0·5 μ. In both cases this material was well mixed with macerated carrot to make it readily acceptable. The food was seen to pass rapidly into the crop and remained there for twenty to thirty-five minutes before some passed into the stomach. When it reached the stomach the barium sulphate always passed quickly into the intestine, while only a little Thorotrast went directly into the intestine, the majority entering the digestive glands almost immediately and

Food, feeding and digestion

spread rapidly through them. Material which entered the intestine passed through and was voided about eight hours after the meal had been consumed. The passage of the faeces through the intestine therefore takes about seven and a half hours. As faeces were only very rarely observed within the rectum it is likely that they pass very rapidly through this region. Some Thorotrast was still present in the crop and digestive gland twenty-four hours after feeding. As barium sulphate and thorium dioxide are inert materials the explanation of their different behaviour must lie in their particle size.

Several workers have studied the controversial question of whether intracellular digestion of particulate material occurs in the digestive glands of pulmonates (Rosen 1952). It can be concluded that the size of the particles and also the nature of the accompanying material are important. Thus, using carbon black of particle size $0.1\,\mu$, Rosen found in *Helix pomatia* that none of the carbon passed into the digestive gland when mixed with flour; whereas it did when mixed with lettuce or carrot. In other experiments some large particles, pigeons' blood corpuscles (the snails would not eat frog or fish blood) and small crystals of the protein edestein were taken up by the digestive gland cells, whereas the more unnatural colloidal gold and Norite (carbon) were not. As two to five days of feeding were necessary in these experiments before the material was visible in the digestive gland cells it is uncertain how important these findings are in an interpretation of normal digestion. Thorotrast has many advantages over the materials used by Rosen, as it is not only visible in the light microscope but also in the electron microscope: and since it is radioactive it can readily be detected by autoradiography. Walker (1969) was able to show that the digestive cells took up particles of Thorotrast, apparently by phagocytosis, into small apical granules which fused to become identical with the green granules (Fig. 29). No particles were ever found within the calcium or excretory cells. Using enzyme histochemistry Walker was also able to show that the hydrolytic enzymes acid phosphatase and esterase were present in these cells. In vertebrate tissues there is often a clear localisation of these enzymes within lysosomes (Hirsch and Cohn 1964) which have been shown to be involved in the intracellular digestion of protein. Rosenbaum and Ditzion (1963) have obtained detailed evidence for the involvement of lysosomes in intracellular digestion in *Helix pomatia*. It therefore appears likely that the intracellular digestion of particulate material and large protein molecules takes place within the digestive gland cells of pulmonates by a lysosome-type mechanism. Also within vertebrate cells carrying out extensive intracellular digestion of this type, large amounts of indigestible material accumulate first as dense bodies then as large masses of pigmented material, lipofuscin. It is possible that the excretory cells represent degenerating digestive cells, replete with this indigestible material.

Intracellular digestion of particulate material has been established in lamellibranchs and some gastropods (Owen 1966, Martoja 1964a). In these cases amoebocytes are very important for the transport of the particles and/or their digestion. In *Agriolimax reticulatus* it is unlikely that amoebocytes are involved in digestion since particulate material entering the stomach passes almost immediately into the digestive glands, much too fast for amoebocyte transport; and in sections no amoebocytes are visible in the lumen of the digestive tract or digestive glands. The quantitative importance of intracellular digestion in pulmonates is unknown but these animals are certainly able to deal with quite large amounts of food in this way.

If the animal is sectioned after a meal of plant material the faeces are seen to contain clearly recognisable pieces of plant tissue. The larger pieces have empty cells on the outside but cell contents are present towards the centre of the piece. *Limax maximus* and *Agriolimax agrestis* are able to utilise only a small percentage of the starch grains in uncooked potato but the majority of grains from cooked potato are digested (Cardot 1924, Crozier and Libby 1925).

This lack of complete breakdown is one of the reasons for the characteristic colour of the faeces. That the plant cell walls do break down is indicated by the presence of empty cells and especially by the easily recognisable and separated spiral thickenings of protoxylem vessels (Evans and Jones 1962a). The plant cell wall is very complex, consisting not only of cellulose fibres but many other complex polysaccharides around and between them. Slugs with their impressive array of polysaccharidases are well equipped to break down this material but presumably, in the absence of trituration, breakdown of large pieces is incomplete.

Some hours after the animal has finished feeding, the crop is almost empty. By the time that the animals become active again towards evening the crop is full of juice. It is therefore possible that the slug digestive gland is concerned with absorption during the digestion of a meal but secretes enzymes during periods of rest.

The uptake of soluble material has also been studied by Walker in *Agriolimax reticulatus*. Carrot mixed with palmitic acid, glucose,

Food, feeding and digestion

galactose, or glycine, all labelled with radioactive isotopes, was fed to the slugs and after suitable time intervals the animals were killed, and sections then autoradiographed. It was found that while all regions absorbed these materials there was maximum uptake of palmitic acid by the crop wall and perhaps maximum uptakes of glucose, galactose and glycine by the digestive gland. Particles of Thorotrast were not taken up by the cells of either the crop or intestine walls. Active transport of material across the intestinal wall has been shown in one mollusc, the chiton *Cryptochiton stelleri* (Lawrence and Lawrence 1967).

We are now in a position to summarise the sequence of events that occur during the digestion of a meal by a herbivorous slug such as *Agriolimax reticulatus*. The radula reduces the food to pieces of very variable size, which pass back into the buccal cavity and move rapidly into the oesophagus. They become well mixed with saliva and mucus but probably very little digestion has occurred by the time they reach the crop. Within the crop a large amount of crop juice is present, contributed largely by the digestive gland but perhaps also by the crop wall. The food material and enzymes of the crop juice circulate around the crop and are mixed by peristaltic contractions of the crop wall. Digestion of fats, proteins and polysaccharides starts, and as the plant cell walls are slowly broken down more digestible material is released. Soluble materials, especially fatty acids, are absorbed through the crop wall. Portions of the food mixed with crop juice are released at intervals into the stomach and here sorting occurs, with soluble and finely particulate material passing into the digestive glands, while the large particles are transported to the intestinal groove where they aggregate to form the faeces. Within the digestive gland cells intracellular digestion of the small particular matter and material of high molecular weight takes place. Indigestible material together with cell debris pass out of the digestive gland in a string of mucus—the liver string—and become incorporated into the faeces. Within the stomach the faeces are loose accumulations of material, but when they pass into the intestine they become more consolidated and are encapsulated in a layer of mucus. As they pass back along the intestine further mucus layers are added so that the faeces form a faecal string, which is voided to the outside. Any soluble material that remains after passage through the stomach is absorbed by the intestinal wall.

While we thus have a reasonable understanding of the events occurring during digestion it should be remembered that this was derived from a study of only two species of slug feeding on plant material. Even in these animals we have little quantitative information, particularly about the importance of digestion and absorption in the various regions. There is practically no information on digestion in the very interesting carnivorous slugs and those that eat fungi and lichens.

4

METABOLISM, RESPIRATION, BREATHING, CIRCULATION, BLOOD, WATER RELATIONS AND EXCRETION

METABOLISM

Energy balance

Of the food eaten by an animal (I), some passes through the digestive system unaltered (E), and some is assimilated (A), i.e. $I = E + A$. Part of the assimilated food is rapidly used to provide energy (via respiration) for locomotion, synthesis and other bodily functions (R), while the remainder is retained as living matter (P), i.e., $A = R + P$. Some of the retained material may be excreted later or lost as reproductive products.

The energy balance has been found for *Arion ater rufus* in the laboratory (Stern 1969). These slugs were maintained throughout their development under constant conditions (15°C, 100% R.H., 12 hr light period). On a diet of lettuce they consumed an average of 78 cal daily. Seventy-four percent of the ingested food was assimilated: a much higher proportion than in other herbivores where values of less than 50% are obtained. This effective utilisation of food implies efficient digestion, even of material such as cellulose. Fifty-four percent of the food ingested was used to provide energy for the various body activities and 20% retained as living matter. The energy balance varied to some extent throughout the growth of the animal, and this was related to reproductive maturation.

Carbohydrate metabolism

A study of the depletion of food reserves during starvation shows that the main food reserve of pulmonates is carbohydrate (Emerson 1967). Glucose is a normal constituent of the blood and in the slug

Ariolimax columbianus Meenakshi and Scheer (1968) report levels of 27·9 ± 6·9 mg glucose/100 ml. This represents 83% of the total blood carbohydrate; of the remainder 3·6 mg/100 ml was present as glycoprotein and 1·8 mg was unidentified. Goddard and Martin (1966) comparing the energy reserves of a wide variety of animals calculate that in *Helix pomatia* the blood sugar can provide sufficient energy for only twenty-four mins, very much less than in vertebrates.

When the animals feed the amount of glucose in the blood rises. Perhaps the most complete data available are those of Holtz and Brand (1940) on *Helix pomatia*. They showed that the normal level of glucose in the blood was 8 mg/100 ml but when they fed the animals on cabbage soaked in glucose this rose to 25 mg/100 ml after one hour, and 49 mg/100 ml after three and a half hours. If feeding was then stopped the blood sugar fell to 22 mg/100 ml after two hours and 17 mg/100 ml after five and a half hours. It is presumed that the glucose is converted into glycogen for storage in the tissues. The amount of glucose in the blood increases for relatively short periods following a meal and then returns to the normal level, suggesting that there must be a process of regulation (Goddard, Nicol and Williams 1964).

In *Ariolimax columbianus,* glycogen is present in large amounts in the digestive gland and the foot (Meenakshi and Scheer 1968) amounting to 3·48% and 2·2% of the wet weight respectively, while only very small amounts are present elsewhere. While glycogen appears to be the main food reserve of the adult, galactogen, a polymer of galactose, is the reserve material for the developing embryo (p. 85). Goddard and Martin (1966) have recently reviewed the somewhat fragmentary work on the biochemistry of the oxidative metabolism of carbohydrates in molluscs, which are difficult animals to study in this respect. Little work has been carried out on slugs. It can be stated however that most, and perhaps all, of the enzymes of the classic Embden–Meyrhof glycolytic cycle and the Krebs citric acid cycle are present, but there appear to be considerable differences in the quantities of the various enzymes present. It is hoped that future studies will further clarify carbohydrate metabolism in slugs, particularly as they are free of the complications of dormancy states.

The synthesis of complex polysaccharides, both galactogen and mucoids, is by way of uridine diphophosphate compounds as in other groups of animals. It is found that radioactive glucose, galactose and maltose are utilised for the synthesis of such compounds (Meenakshi and Scheer 1968).

Metabolism

Protein metabolism

Free amino acids are present in small amounts in the blood of pulmonates, about 12 mg/100 ml blood (Kerkut and Cottrell 1962, Campbell and Speeg 1968a).

Molluscs appear to be able to synthesize most, perhaps all, of their amino acids. Campbell and Speeg (1968a) studying the synthesis of arginine in the snail *Otala* found that radioactive carbon in the form of bicarbonate was incorporated into twenty-six different amino acids, which had been synthesized into protein. The largest amounts of the protein containing these amino acids were found in the digestive gland and the kidney; it is therefore likely that the most rapid synthesis of proteins takes place in these organs.

Lipid metabolism

As a result of the very great improvements in analytical methods for lipids which have been developed in the last ten years, it has been found that molluscs contain a variety of unusual lipids.

The lipid content of *Ariolimax columbianus* is 1·12% of the wet weight, large *Arion ater* 0·8% (equivalent to 8·0% of the dry weight), and small *Arion ater* 2·3% (Thompson and Hananan 1963). In the large *Arion* approximately one-third of the lipid is present in the viscera. Of the fat approximately 20% is triglyceride, 40% phospholipid and 40% steroid (mainly cholesterol and its esters). Little difference was found in the distribution of phospholipids, but triglycerides are found mostly in the body wall and foot while the steroids are mainly present in the viscera. Analysis of the triglyceride fraction revealed the existence of a high percentage of glyceryl ether derivatives at concentrations previously found only in the oils of elasmobranch fish. The glyceryl ether content of the phospholipids is the highest known for any animal. It is also unusual in that only one or two glyceryl ethers are present and these are saturated. Another unusual phosphorus-containing lipid has been found by Liang and Rosenberg (1968) in *Lehmannia poirieri* and *Limax flavus*. They discovered that 2-aminoethylphosphonate (this is unusual in that it contains a carbon-phosphorous link), probably formed from a citric acid cycle intermediate, was incorporated into lipids and an insoluble fraction.

The phospholipids of slugs are therefore unusual and it is to be hoped that further biochemical studies will provide an explanation for this. It is assumed that the phospholipids are largely involved in membrane structure, while most of the neutral lipid is probably involved in metabolism. Indeed, electron microscopy and histochemistry reveal the presence of much lipid in the cytoplasm of a range of cell types, notably those in the digestive system (Walker 1969) and gonad. A complex lipid mixture is also secreted on to the outside of the eggs of some slugs (Smith 1966 and Bayne 1968).

Steroids are of great interest because of their possible role as hormones. Cholesterol, however, accounts for approximately 30% of the total steroids. The remaining steroids consist of unidentified cholesterol esters but also 7-deshydrocholesterol. Voogt (1967) was able to show that sterols were readily synthesized from acetate in the slug tissues so that they do not depend on plants for the supply of these compounds. Gottfried and co-workers (1966 and 1967) have found that the eggs and parts of the reproductive tract have very considerable synthetic ability, synthesizing various steroids from endogenous substrates. The significance of these findings is as yet uncertain.

RESPIRATION

All the various activities of an animal lead to the consumption of oxygen and production of carbon dioxide. The measurement of oxygen uptake is the most frequently used measure of an animal's metabolism, but it is often difficult to relate the measurements made by various workers. Respiratory rate is affected by: temperature, both during and before the experiment; size; age; and activity, both locomotory and metabolic (e.g., digestion).

Temperature has a profound effect on body processes, but the way in which an animal responds to changes in temperature depends on the temperature history of that animal, both in the long and in the short term. It is often found that animals collected from the extremes of their geographical range respire at different rates when measured at one temperature. This is due to temperature *adaptation* (Hart 1957) and it has a genetic basis. Variations in respiratory rates are also found in the same animals at different times of year and this is termed temperature *acclimatisation*. Animals can be kept in the laboratory under constant conditions and one factor can be varied at a time, the effects of such variations being termed *acclimation*. Such experiments show two effects of temperature change. Firstly, immediately after the change, there is a very rapid alteration of the respiratory rate which over a period of hours is reduced until a new rate is adopted. Secondly over a longer period this new rate gradually changes until a stable rate results. In *Helix pomatia* it is found that the short term acclimation takes ten to fifteen hours, while the longer term acclimation takes fourteen days or longer (Blažka 1955). An increase of temperature from 15 to 25°C leads initially to an increase of respiratory rate of approximately 100% which after one day has decreased to only 25%, while after fourteen days it has fallen to only 12·5%. If the temperature change is reversed then the respiratory changes are reversed. Short-term acclimation in *Limax maximus* takes about seven days and in *Philomycus carolinianus* twenty days (Rising and Armitage 1969). In some animals the final respiratory rate at the changed temperature may be identical with that before the change of temperature. This has been found for *Milax budapestensis, Agriolimax reticulatus* and *Arion hortensis,* when in the resting state, by Newell (personal communication).

The changes of respiratory rate reflect changes in the metabolism of the animal's tissues. Thus, when animals which have been reared for a long period (months) at a low temperature are compared with those reared at a high temperature it is found that the respiratory rates are always higher for the cold-acclimated animals. Thus Roy (1963) found that when *Arion fasciatus* acclimated at a wide variety of temperatures were compared, there was a 1–1·5% decrease in metabolic rate, as measured by oxygen consumption or heat production, for every 1°C increase in acclimation temperature. Similar results were obtained for *Limax maximus* and *Philomycus carolinianus* (Rising and Armitage 1969) although animals collected at different times of year showed some differences in their responses to temperature change. When temperature-acclimated *Limax maximus* were placed in a temperature gradient their preferred temperature was at, or slightly above, their acclimation temperature. When *Limax maximus* which had been acclimated at 25°C were placed in the 5–9°C region of the temperature gradient they remained immobile. By contrast the 5°C acclimated slugs were active at 5–9°C.

Slugs are therefore able to compensate temperature changes, and are thus within certain limits independent of the environmental temperature.

In *Helix pomatia* (Kerkut and Laverack 1957) the sum of the oxygen uptakes of the isolated organs exceeded that of the whole animal by up to 50%. The respiratory rates of the brain and digestive gland were the highest and the foot the lowest. Nopp and Farahat (1957) found that these isolated tissues also varied in their responses to temperature change. It must be concluded that the level of metabolism of the whole animal and indeed of individual tissues is controlled by a mechanism probably neural or endocrine which is as yet undefined.

With an increase in the activity of an animal, particularly by locomotion, oxygen requirements increase. Newell (personal communication) has shown that with slugs, there can be an increase in oxygen consumption of up to 400%, but the increase was highly dependent on the temperature. While the resting slugs showed almost no change in respiratory rate with alteration of temperature, a very marked change occurred when the animals were active. In chemical systems, as the temperature is raised by ten degrees the reaction rate increases by approximately two, which is usually stated as $Q_{10}=2$. In his resting slugs, Newell found a Q_{10} of about 1, while in the active slugs it varied from 1·7 to 2.

In studies on metabolic rate it is usually only the oxygen consumption that is measured, because the amount of carbon dioxide produced is dependent on the type of food material being metabolised. But as this ratio of carbon dioxide production to oxygen consumption, termed the respiratory quotient (RQ), is characteristic of the type of food substrate, this can provide useful information. The RQ for carbohydrate metabolism is approximately 1·0 while that for protein is 0·8 and fat 0·7. The values reported for *Arion fasciatus* and *Agriolimax agrestis* are approximately 1·0 so that in these animals the respiratory substrate is likely to be carbohydrate.

BREATHING

The lung surface covers the roof of the pulmonary cavity but in some species extends on to its walls and floor. The lung surface areas of a number of molluscs varies from 7 to 9·4 cm^2/g wet body wt (Pelseneer 1935). Large pulmonary veins drain into the auricle. The veins branch extensively and there are many anastomoses between them. The veins have the same structure as the arterioles present elsewhere—there is an inner incomplete endothelium lying on a layer of collagen fibres and an outer layer of muscle fibres lacking a consistent orientation (Pohunkova 1967). Over the outer surface of the veins there is a respiratory epithelium of cubical or squamous epithelial cells bearing short microvilli, while between the vessels and opening through the epithelium there are numerous goblet mucus cells. The mucus secretion maintains a continuous layer of liquid over the epithelium. The distance between the wall of the veins and the lumen of the lung varies between 6 and 10 μ. If it is assumed that the tissue has similar properties to vertebrate connective tissue it can be calculated that in *Agriolimax agrestis* the diffusion potential for oxygen through the lung surface is six times greater than the animal's total oxygen requirements. Carbon dioxide diffuses at least twenty times more readily than oxygen and so its rate of diffusion is unlikely to be a limiting factor.

Oxygen that has diffused through the lung wall has to be transported and distributed by the blood. Using information derived from a number of pulmonates it is calculated that the blood can transport 1 μl oxygen/min/g bodywt—this is one third of the requirement for respiration! It thus appears probable that skin respiration is extremely important to slugs. Little work has been carried out on the importance of the body wall of slugs for respiratory exchange. It was however shown by Schuurmans-Stekhoven (1920) that if the lung of *Agriolimax agrestis* is filled with paraffin, 56% of the normal oxygen uptake can still take place through the body wall.

Various mechanisms to promote the exchange of gas within the lung have been reported (Ghiretti 1966a). Under normal conditions the pneumostome is open for the majority of the time and the floor of the lung can then be seen to move up and down. This is particularly clear in the case of *Arion ater* which has a very large opening. This movement pumps air in and out of the lung. In *Helix* forced ventilation is also used as follows: air is drawn into the lung by lowering the floor of the pulmonary cavity; the pneumostome is closed and the floor raised, thereby increasing the pressure; opening of the pneumostome then allows exit of the air. These two breathing mechanisms do not appear to be stimulated by carbon dioxide; but lowering of the oxygen content of the air below 11%, or raising the temperature, produces marked increase in respiratory movements.

BLOOD SYSTEM

The blood vessels can be readily studied by injecting them with liquids such as Indian ink or, if a permanent specimen is required, rubber latex. In order to demonstrate the whole system several animals must be injected. A diagram of the blood system of *Agriolimax reticulatus* is shown in Fig. 30.

At the end of the capillaries blood leaks out into the haemocoel or into one of the many blood sinuses. In *Arion* Jourdain (1879) described funnel-shaped openings at the ends of the capillaries while

C

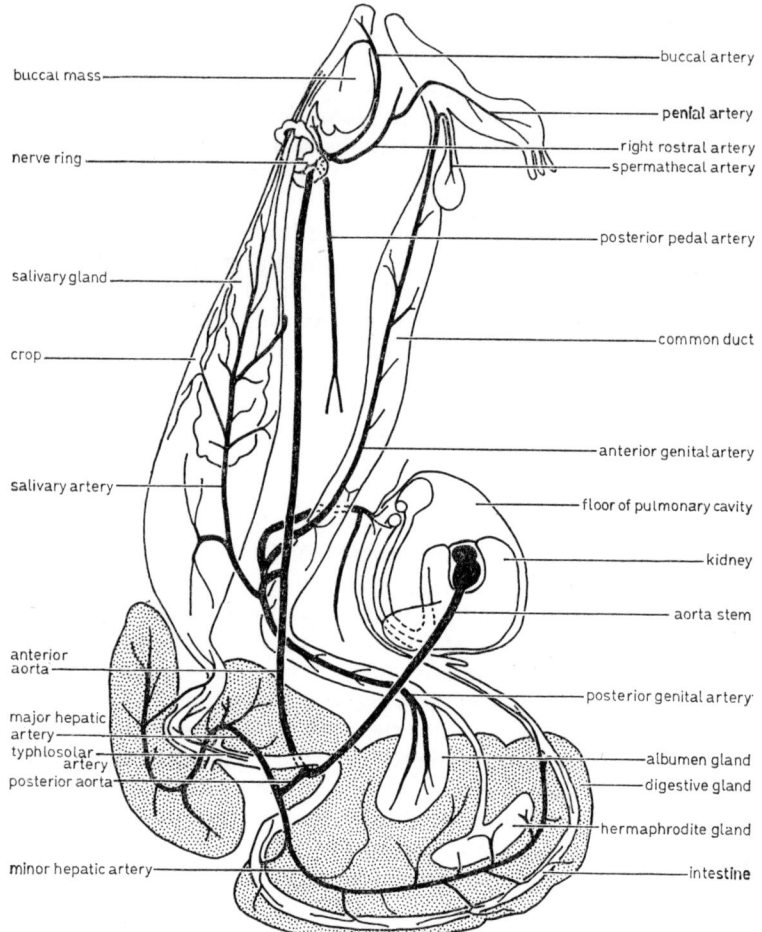

Fig. 30 *Agriolimax reticulatus* arterial system (after Laryea 1970, Walker 1969, Garner 1970).

in the basommatophoran snail *Lymnaea stagnalis* there are muscular sphincters (ostioles) around the openings (Carriker 1946). In *Agriolimax reticulatus* the capillaries open without any apparent specialisations (Laryea 1970).

The various blood sinuses of the foot and body wall join around the lung and kidney to form a number of interconnecting sinuses

Blood system

approximating to a ring vessel. From this vessel there are branches to the lung and kidney. Blood from the kidney and lung is collected into veins which connect to the auricle. The walls of the sinuses and veins appear to have very little organised structure. There is considerable variation in the number and arrangement of the sinuses, apparently because there are many interconnections.

The heart consists of a mass of muscle fibres which form a meshwork. The auricle wall is quite thin whereas the ventricle wall consists of a thick mass of muscle fibres, many crossing the lumen. Externally the heart is covered by a thin epithelial layer, or epicardium, which is continuous with a similar epithelium lining the pericardium. Internally an epithelial lining (endocardium) is absent so that the muscle cells are in direct contact with the blood. At the atrio-ventricular junction there is a ring of dense connective tissue, in which are embedded the ends of the muscle cells. When examined in the electron microscope (Nisbet and Plummer 1966, 1968) the heart muscle cells can be seen to contain thick and thin filaments as in vertebrate muscle but they are less well organised. Depending on the arrangement of the filaments the muscles may appear striated, spirally striated, or unstriated. The abundant mitochondria occur as a solid core running down the centre of the muscle cell with only a few along the edges. Cells are joined end to end by means of many specialised adhesive areas or desmosomes.

Simultaneous contraction of the ventricular muscle fibres propels blood into the aorta. A flap valve at the base of the aorta prevents back pressure returning the blood. Blood then passes along the arteries through the capillaries and into the venous sinuses to return via the veins of the lung and kidney. The contraction of the ventricle in the fluid-filled and inelastic pericardium results in an expansion of the thin-walled auricle, thus drawing in the blood from the veins (Krigjsman and Divaris 1955). No valve is present between the auricle and the veins so that it is perhaps the general body pressure which prevents too much backflow of blood into the veins when the auricle contracts. Contraction of the auricle forces blood into the ventricle, and backflow of blood is prevented by the auriculo-ventricular valve.

Slugs have a higher frequency of heart beat than other molluscs of similar weight (Schwartzkopff 1956). Thus *Agriolimax agrestis* of body weight 0·213g has a heart beat of 90·6/min, and *Arion ater* of 16·6g one of 48·8/min. Animals of a single species show a decrease in the frequency of heart beat with increase in body weight.

In *Helix* (Schwartzkopff 1954) the pressure in the cannulated aorta reaches 35 cm water; its output is 5 ml/100 g body wt/min (man: 6–10 ml/100 g body wt/min); and the work done by the heart is 3,600 g cm/min/g heart wt (man: 4,000 g cm/min/g heart wt). Calculations by Chapman (1967) show that it takes approximately 6 min for the blood of a 10 g *Helix pomatia* to be circulated around the body (21 sec in man).

There is now general agreement that the pulmonate heart is myogenic (i.e. it will continue beating when removed from the body and in the absence of any innervation) and that there is no localised pacemaker. Stretching of the heart whether by pulling or distension is required, however, for the beat to be regular and at the normal rate. The heart can be slowed or accelerated by nervous stimuli. Acetyl choline has been shown to inhibit the heart beat, while 5-hydroxytryptamine accelerates it. These neurohumours appear to be released by the nerves controlling the heart beat (Hill and Welsh 1966). Cottrell and Osborne (1969) have recently described a collection of neurosecretory neurones at the auriculo-ventricular junction but their function is not yet known.

Comparatively little is known about the structure of pulmonate blood vessels. In the walls of the large arteries of *Agriolimax reticulatus* there is a diffuse layer rich in collagen fibres. Outside of this there are inner circular and outer longitudinal layers of smooth muscle with fibroblasts, nerve axons and collagen fibres between the cells. Scattered cells with much branched processes are present on the luminal surface of the collagenous layer. In spite of all the processes the cells cover only between 15 and 50% of the surface. These cells, closely resembling other cells present throughout the connective tissues, are probably a form of wandering amoebocyte. The small arteries have a less ordered arrangement. It appears therefore that there is no liquid-tight lining to the vessels, as in vertebrates, and fluid must leak readily through the walls of the vessels.

In *Agriolimax reticulatus* nerve axons were found to terminate on the muscle cells in the walls of the arteries and arterioles (Laryea 1970). While a rich innervation of the contractile elements of the arterial walls has been shown, nothing has been discovered of the physiology of the vessels. Anastomoses (interconnections) between arterioles and capillaries are common, so blood is possibly shunted to areas where the demand is greatest. When traceable materials are injected into the blood system, vessels are only rarely equally well filled, and whole regions may not receive any of the injected material. Thus, local constrictions of the lumen seem possible.

Amoebocytes

Within the blood of pulmonates there are white cells, or amoebocytes, which have varying forms but are probably variations of one cell type. No figures are available for the number present in the blood of slugs, but in the prosobranch *Bullia* there are 3,800–7,200/mm^3 (Brown and Brown 1956) and in the basommatophoran *Lymnaea stagnalis* 50,000–250,000/mm^3 (Müller 1956). Amoeboid cells of varying kinds are found throughout the tissues of slugs and may contain granules of food reserve materials. It is possible that amoebocytes play an important part in the distribution of food material to the organs.

Many workers have shown that if particulate material is injected into the blood system of pulmonates it is rapidly taken up by amoebocytes. Brown (1967) showed that Thorotrast particles were phagocytosed in *Helix* by amoebocytes which then migrated through the tissue to the lumina of the lower female tract, the mantle edge, and mid-gut where they were voided. In their previous study of the prosobranch *Bullia,* Brown and Brown (1965) were able to show that injection of foreign material gave rise to a rapid multiplication of amoebocytes so that they reached concentrations in the blood of up to 23,000/mm^3. After injection the number of mitotic cells rose to a maximum of eighteen in a thousand contrasting to the normal state of less than one in a thousand. The very great increase in cell number was due not only to divisions of pre-existing blood cells but also to the detachment and multiplication of cells lining the blood vessels and those in the connective tissues. When the kidney of *Agriolimax reticulatus* is infected with the protozoan parasite *Tetrahymena rostrata,* rapid multiplication of amoebocytes in the veins and connective tissue of the kidney occurs around the site of infection (Brooks 1966). As the infection progresses, vessels in the lung and the cells lining the heart are involved. When the parasites reach the connective tissue they sometimes become surrounded by cells which form a cyst. Often so many amoebocytes are present that they form masses blocking the vessels. Similar rapid multiplication of amoebocytes has been recorded following mechanical injuries (Chétail 1963).

In recent reviews of the defensive mechanisms of molluscs (Feng 1967, Acton, Evans and Bennett 1969), it is concluded that amoebocytes are the major if not the only defence mechanism against infection. Attempts to produce antibodies by injection of bacteria or viruses failed. Mollusc blood contains haemagglutinins, which precipitate rabbit blood, but no normal function for this property has been found. There is evidence that there are immune reactions against helminths in the aquatic snail *Australorbis* (Michelson 1964) but much work is required to discover if other systems of this type exist.

Leydig cells

In some slugs the arteries supplying the organs, particularly the digestive gland, stand out clearly, because of their white appearance. This is due to a close association of Leydig cells with the walls of the arteries. These cells, which are large and contain one or many vacuoles or granules giving strong histochemical reactions for calcium salts, are also found in connective tissues elsewhere. Some materials (e.g., haematoxylin), when injected into the haemocoel, are selectively absorbed by these cells and apparently broken down within them (Filhol 1938). Filhol concluded that one of the main functions of these cells is excretion by accumulation, but there is also clear evidence that they store glycogen and calcium. Much work is needed to unravel the functions of these obviously important cells.

Arterial gland

In *Agriolimax reticulatus,* Laryea (1969) found that on the walls of the cephalic artery and its branches, but most extensively along the posterior pedal artery, there are groups of another type of cell which lacks the calcium of the Leydig cell. There is considerable variation in the amount present and the distribution of this tissue, which Laryea terms 'the arterial gland'. The cells are irregular with intracellular canals and contain many protein secretion granules which are discharged into intracellular ducts communicating with discrete channels between groups of the cells. Over the surface of the gland there is a thin layer of collagen fibres (less than 1 μ thick) separating the gland from the haemocoel. The secretion is apparently discharged through this membrane into the blood. The gland cells are rich in copper, so it was conjectured that the gland might be involved in haemocyanin formation. Attempts to prove that the protein of the granules is a haemocyanin precursor have so far proved unsuccessful. The variation in the size of the gland appears to be unrelated to any body process, including reproduction; but following the end of egg laying, when the animal has only a short while to live, the arterial gland is found to be disintegrating. The function of this interesting gland requires further study.

Blood volume

The soft slimy nature of slugs clearly indicates that a large amount of water is present. In *Arion ater* (Martin *et al.* 1958) the water content of the body is $86.3 \pm 3.8\%$. Such large amounts of water are characteristic of most molluscs. These authors also determined the blood volume by injecting a known amount of inulin and determining its eventual dilution. Of the total body weight $36.6 \pm 10.4\%$ was found to be blood.

WATER RELATIONS

Slugs keep their surfaces moist and since they have no impermeable covering their water content fluctuates greatly. Howes and Wells (1934) found that when individual *Arion ater* or *Limax flavus* were kept under fairly constant environmental conditions they showed periodic changes in body weight, amounting to 15–30%, over periods of five to ten days. These fluctuations were unrelated to food intake and occurred only when the slugs were kept in conditions of fairly high humidity. The fluctuations were therefore due to variations in water content. It was also noted that there was some tendency for hydrated animals to be active while dehydrated animals were inactive.

When the snails *Mesodon thyroidus* and *Allogona profunda* are handled (Blinn 1964) they squirt out a watery liquid, the amount of which is directly related to a weight cycle. The watery fluid is emitted from the pneumostome, and as the animals have transparent shells it is possible to see that the fluid is stored in the pulmonary cavity. This fluid, termed pallial water, appears to be fairly pure water in these animals but it is viscous in other species. Blinn showed experimentally that the pallial water originates from water drunk by the animals which passes across the gut wall, and eventually into the lung cavity. The pallial water appears to act as a reservoir reducing fluctuations in the water content of the body itself. Blinn refers to older observations by Künkel which indicate that pallial water may similarly account for part of the weight fluctuations in slugs, possibly as much as 50%.

Dainton (1954a) records that she has never seen slugs drinking, but that they very readily take up water through the skin. Loss of water must occur either by evaporation from the body surface or by secretion of mucus for locomotion. Evaporation from the surfaces of *Agriolimax reticulatus, Arion subfuscus, Arion ater, Limax maximus* and *Milax sowerbyi,* maintained at 45% R.H., was 3–5% of the body wt per hr. When crawling in a saturated atmosphere, so that losses by evaporation were minimal, *Agriolimax reticulatus, Limax maximus* and *Arion ater* lost 17% of their body weight in 40 minutes. Animals became less active with progressive loss of water, so that increased mechanical stimulation was needed to make them crawl. When they had lost about 25% of their body weight (after about three hours and much stimulation) they became non-reactive. As a loss of 17% of body weight, whether by evaporation or by locomotion, always resulted in a similar effect on behaviour, it is likely that water rather than loss of mucus is the important factor. In order to discover how water was taken up, Dainton stimulated a *Limax maximus* until its body weight had been reduced from 9·1 g to 6·6 g, then she suspended it above, but completely out of contact with, a water surface in a sealed container. Within two hours its weight was back to normal and it regained its activity. As slugs normally live in an almost saturated environment, if not actually in contact with free water, this attribute must be very important. When slugs crawl actively in a non-saturated atmosphere they rapidly lose water and die.

Slugs do, however, have the ability to survive losses of large amounts of water. Dainton mentions that *Agriolimax reticulatus, Limax maximus* and *Arion ater* always survive losses of 50% body weight, regaining their normal weight in a few hours when placed on a damp surface. Howes and Wells quote Künkel's description of a *Limax tenellus* that recovered after losing 80% of its weight.

These experiments clearly show that slugs have very little physiological control over their water loss. In *Helix aspersa,* Machin (1965) has found that, while there is no control over evaporation losses from the body wall, the mantle rim, which closes the shell opening when the animal withdraws, is capable of regulating water loss. Machin also found that the skin of *Helix* is normally kept moist by secretion of mucus, so that evaporative losses are also related to mucus secretion. Isolated pieces of skin dried up, but under experimental conditions, when saline at a pressure of 10 cm water was present on the inner surface, the mucus cells extruded sufficient mucus to keep the skin moist. The pressure of blood in the haemocoel appears to vary between 7·5 and 25 cm water (Schwartzkopff 1954). When animals were placed in strong wind currents, mucus flow was increased to keep the skin moist and this was associated with muscular undulations of the body wall. As in other pulmonates, the skin is warty, and it appeared that mucus was secreted into the grooves between the warts and these areas acted as a sort of reservoir. When mucus was needed it was moved up on to the surface of the warts probably by ciliary action. The rate of water loss by evaporation was also influenced by the body shape so that loss was at a maximum when the animal faced into the wind, losses were 83% of this when at right angles to the wind, and 68% when facing down wind. The shape of the body directly affected the pattern of laminar flow of air over the body, hence the variations in rates of water loss. While work of this kind has not been carried out on slugs, Dainton (1954b) found that *Agriolimax reticulatus* and *Limax maximus*, when stimulated by air currents, increased their speed of crawling when facing down wind, and when facing any other direction they turned to face down wind. Thus, if Machin's findings are applicable to slugs, they minimise evaporative water loss by this behaviour when crawling.

BLOOD COMPOSITION

The fluctuations in weight are due not only to changes in the volume of the pallial water but also to changes in the amount of water in the blood and tissues. It is well established that the blood of terrestrial molluscs is extremely variable in concentration. Thus the freezing point depression (a measure of the total solute concentration) varies from -0.18 to $-0.43°C$ in *Arion ater* (Roach 1963), -0.585 to $-0.715°C$ in field-collected and 0.705 to $-0.820°C$ in laboratory-reared *Agriolimax reticulatus* (Bailey 1970). These fluctuations have been analysed extensively in *Helix* species by Burton (1964). The depression of the freezing point of *Helix* blood varied from -0.24 to $-0.62°C$ and the solute concentration from 14 to 101 g/kg water. There was some regulation of potassium and calcium ions; when the amount of potassium was artificially increased, there was a compensatory increase in the amount of calcium. Although the food (plant material) was rich in potassium there was little change in the blood following a meal. Burton suggests that there are two mechanisms controlling the concentrations of sodium and magnesium, the first coming into operation at one blood concentration to move salts and water from the blood to the tissues, and the other at a rather different concentration moving ions and water from the tissues to the blood. He suggests that since the critical concentrations are widely separated this prevents the mechanisms from being in continual operation thus obviating the possible loss of ions. The inorganic constituents of *Arion ater* blood are shown in Table 5.

TABLE 5
THE CONCENTRATION OF IONS IN THE BLOOD OF ARION ATER

Ion mM/kg water	SPECIES		
	Arion ater ater[1]	Arion ater rufus[1]	Arion ater rufus[2]
Sodium	53 ± 5	60 ± 9	61·5 ± 5·72
Potassium	—	2·7 ± 0·4	2·68 ± 0·80
Calcium	5·6 ± 1·0	6·9 ± 1·4	2·81 ± 0·48 (20% organically bound)
Magnesium	5·4 ± 0·7	5·4 ± 0·5	5·68 ± 0·87
Chloride	—	—	52·45 ± 1·52
Sulphate	—	—	2·21 ± 0·24
Phosphate	—	—	0·157 ± 0·055
Bicarbonate	—	—	24·81 ± 2·28
Total electrolytes			77·02 ± 8·19

[1] From Burton 1968
[2] From Roach 1963

Roach (1963) showed that the protein constituents of the blood amounted to $10·55 \pm 3·02$ g/l and this was largely haemocyanin. Haemocyanin has a very high molecular weight and is easily removed from the blood by centrifugation at 50,000 g. After its removal some protein still remains in solution. In *Agriolimax reticulatus*, Bailey (1970) has found that this protein fraction constitutes approximately 4% of the total blood protein. Electrophoresis on polyacrylamide gel showed the presence of up to ten constituents in it. The function of these blood proteins is completely unknown.

Blood pH

The blood of pulmonates is alkaline but there are considerable discrepancies between the pH values reported in the literature. Thus Roach (1963) reported a pH of 8·89 for the blood of *Arion ater* and Burton (1969) pH 7·8 for *Helix*. Burton has shown that, in *Helix*, 94% of the buffering capacity of the blood is due to bicarbonate. When this blood is exposed to the air it rapidly loses carbon dioxide by breakdown of bicarbonate (Speeg and Campbell 1968b). Because of this the blood pH changed from 7·4 to 8·8 over a period of two hours. It is therefore essential to measure the pH of pulmonate blood *in situ* with a microelectrode, or use blood collected under oil, to obtain a meaningful value. Using the first of these methods Bailey (1970) has found that the blood pH of *Agriolimax reticulatus, Limax maximus* and *Arion ater*, when measured in the haemocoel between the viscera, was pH 7·70 to 7·75.

Haemocyanin

Haemocyanin is a copper protein of very high molecular weight. Values as high as 8·9 million have been determined for *Helix* haemocyanin (Ghiretti 1966b). These molecules are readily detectable by the electron microscope, using negative contrast techniques, as short thick-walled tubes (300 × 335 Å) consisting of tightly packed subunits (Bruggen, Wiebenga and Gruber 1962). Haemocyanin is easily broken down to smaller units by changes in pH and salt concentrations. From a study of the behaviour of haemocyanin following electrophoresis, ultra-centrifugation and treatment with salts, Lontie *et al.* (1962) conclude that there are two forms of haemocyanin, α and β.

There is evidence that there is a total of 180 subunits in the molecule (Burton 1965a; Wood, Salisbury and Bannister 1968) and that each subunit has two copper atoms attached to the side chains of histidine amino acids. Oxygen is readily taken up by the haemocyanin, one atom being combined with two copper atoms. As oxygen is absorbed the blood changes from the colourless deoxygenated state to the opalescent blue of the oxygenated blood.

If blood is collected from the heart it is clearly blue in colour, while the blood collected from the veins is almost colourless. Thus the blood takes up oxygen in the lung and loses it in the tissues (Spoek, Bakker and Wolvekamp 1964). The maximum oxygen-carrying capacity of the blood is 1·15–2·2 ml of oxygen/100 ml blood—distilled water under the same conditions dissolves 0·7 ml oxygen (Prosser and Brown 1961). It is not clear how saturated the blood normally becomes. As already mentioned carbon dioxide is carried in the blood mainly in the form of bicarbonate in equilibrium with a low concentration of free carbon dioxide and carbonate. When blood reaches the lungs, the carbonic anhydrase, which is present in large amounts in lung tissue (Speeg and Campbell 1968b), causes a change in the equilibrium so that carbon dioxide is released and diffuses out to the lung lumen. The amount of bicarbonate is reduced and so the pH of the blood becomes more alkaline. There is a difference of 0·12 pH units between the blood in the haemocoel and that in the heart of *Helix pomatia* (Burton 1969), which represents a loss of 1·4 ml carbon dioxide/100 ml of blood. This is almost certainly an underestimate of the extent of the change in the lung, as blood entering the heart is of two kinds, some from the lung and some from the kidney.

In *Agriolimax* haemocyanin, there is a very clear Bohr effect, i.e., the oxygen affinity of the blood shows an increase as the blood becomes more alkaline (Fig. 31). Thus at pH 8·78 the blood is 50%

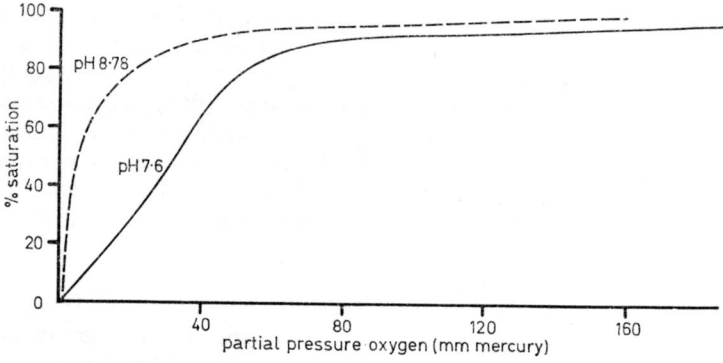

Fig. 31 *Agriolimax*. Effect of pH on the oxygen saturation of the blood (after Prosser and Brown 1961).

saturated when in equilibrium with oxygen at a partial pressure of only 3 mm mercury, while at pH 7·60 to produce 50% saturation of the blood requires an oxygen partial pressure of 32 mm mercury. In the lung the alkaline blood readily absorbs oxygen, while in the tissues where carbon dioxide diffuses into the blood and it becomes more acid, oxygen is readily given up.

Normally associated with a Bohr effect is the Haldane effect, i.e., the haemocyanin releases hydrogen ions as it is oxygenated. Burton (1969) has demonstrated its existence in *Helix* haemocyanin. Because of the Haldane effect, carbon dioxide is more readily lost from the blood in the lung and more readily absorbed in the tissues. Burton (1969) has shown that even without this Haldane effect the weak buffering ability of haemocyanin (4% of the total blood buffering capacity) enables the blood to carry more than twice as much carbon dioxide as haemocyanin free blood.

Where haemocyanin is formed in the body is unknown, as is its metabolism. As the digestive glands and arterial gland are rich in copper they may represent sites of synthesis.

EXCRETION

The kidney of *Agriolimax reticulatus* will be described as we know most about its structure and physiology in this slug (Garner 1970). The kidney bulges into the cavity of the lung from its posterior wall (Fig. 32). It is flattened and approximately circular in outline with a large right posterolateral lobe. The heart and pericardium occupy a notch in its anterior edge. A thin-walled and flattened primary ureter runs over the surface of the right side of the kidney. It continues as the secondary ureter which is closely applied to the rectum. In *Agriolimax* and other members of the Limacidae the secondary ureter discharges into a bladder which has a slit-like duct to the exterior (Fig. 32 and 33). A flap valve where the ureter enters the bladder presumably prevents the passage of material from the bladder back to the ureter.

There are many large folds on the walls of the kidney which project into its lumen. The large venous supply enters the kidney at the tip of the posterolateral lobe and breaks up into smaller vessels which pass into the core of the folds. At the anterior end of the kidney the blood vessels discharge into the large kidney vein, which opens into the auricle.

A continuous epithelium of cubical or columnar cells, the nephrocytes, lines the whole inner surface of the kidney. Underlying the epithelium there is a sparse layer of connective tissue which is bathed by the blood. Characteristically, the nephrocytes contain a large concretion within a vacuole. From a study of sections taken from several animals a sequence for the development of the large vacuole can be suggested. Several vacuoles in the upper cytoplasm fuse to

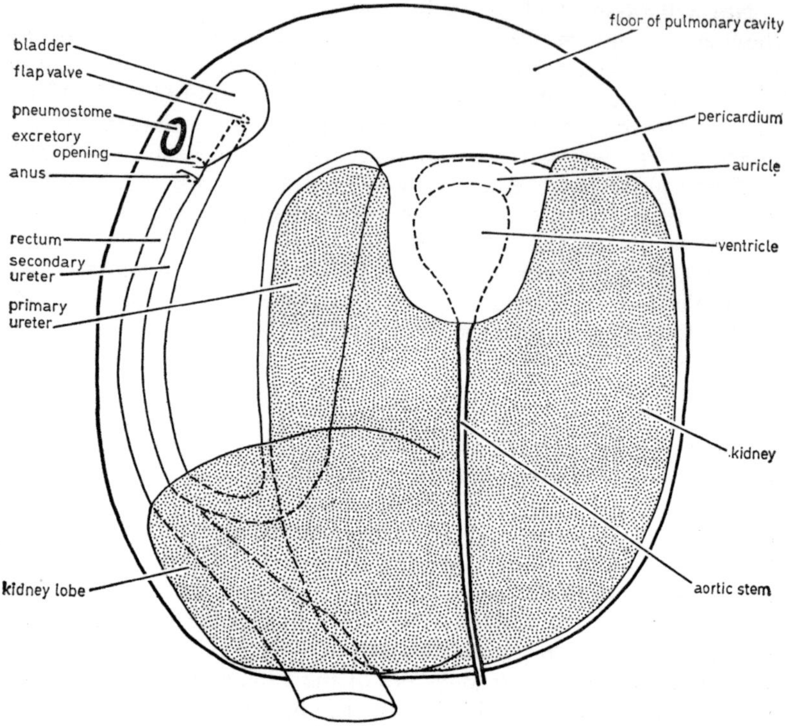

Fig. 32 *Agriolimax reticulatus* heart, kidney and associated organs as viewed from beneath the floor of the pulmonary cavity (after Garner 1970).

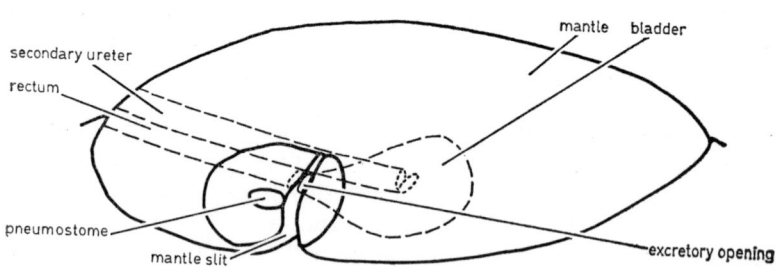

Fig. 33 *Agriolimax reticulatus* mantle and openings (after Garner 1970).

Excretion

form larger vacuoles until a single large vacuole results. Small vacuoles are usually found associated with this large vacuole so that material may continually be discharged into it. The contents appear to crystallise from the outside until the vacuole is almost filled by a large concretion. It is not clear how the concretion is discharged, but in some cells the upper region appears to pinch off and many cytoplasmic fragments are found in the ureter. The upper surface of the cell has a layer of microvilli, and there are many complex areas of intercellular adhesion.

In the primary ureter the walls are much folded. In the electron microscope it is seen that the cells on the walls of the folds interdigitate extensively with neighbouring cells. Many mitochondria occur in the interdigitating cell processes. The luminal surface of the cell has a very well-developed brush border and the upper quarter of the cell is almost filled with non-staining vacuoles. The cells on the summits of the folds are ciliated and lack the extensive interdigitations. It is possible that these ciliated cells are also secretory.

The secondary ureter has a very well-developed smooth muscle coat. The lining epithelial cells have a brush border but they are characterised by the great density of their cytoplasm, large intercellular spaces and intracellular ducts which are flanked by numerous golgi bodies. The cytoplasm of these cells contains masses of glycogen and many large mitochondria with tightly packed parallel cristae.

The excreta of slugs has not been studied as extensively as those of snails. *Agriolimax reticulatus* (Garner 1970) usually deposits one string of largely solid excreta (amounting to 0·5mg) in 24 hours. This deposition occurs in the early part of the evening. The number and size of concretions in the nephrocytes and the amount of uric acid in the kidney is at a minimum at 6 p.m. and increases thereafter to a maximum at 10 a.m. It appears that when the concretions reach their maximum size they are released and stored in the ureter until deposition. This apparent diurnal rhythm of excretion contrasts with the long periods between deposition of excreta in *Helix* (Vorwohl 1961).

Purines (uric acid and xanthine) account for 92% of the nitrogen in the excreta and there are also traces of hypoxanthine. Comparisons with the excreta of active *Helix* and *Otala* are shown in Table 6. As Speeg and Campbell (1968a) have suggested, the term 'uricotelic', usually given to organisms excreting mainly uric acid, is inappropriate for these animals and 'purinotelic' would be better. The results of earlier workers on the excreta of molluscs are inaccurate, chiefly

TABLE 6

PURINE CONTENT OF THE KIDNEY

	μM/g Kidney		
	Uric acid	Xanthine	Guanine
Helix pomatia[1]	504	575	139
Otala lactea[2]	808	190	194
Agriolimax reticulatus[3]	467	187	Absent

[1] Jezewska, Gorzkowski and Heller 1963
[2] Speeg and Campbell 1968a
[3] Garner 1970

because they did not recognise that bacteria will decompose the uric acid quite rapidly to form urea and sometimes ammonia; in addition, the presence of other purines was not taken into account. The usual statement that *Arion* excretes large amounts of urea requires confirmation as urea has not been detected in the excreta of *Agriolimax reticulatus*. Speeg and Campbell (1968b) have shown that *Helix* and *Otala* both excrete gaseous ammonia, particularly during periods of dormancy, when it can account for almost 30% of nitrogen excretion. These authors have suggested that ammonia is formed from urea by the action of urease; it is transported by the blood; and it is released at the surface of the lungs and through the shell. While urease has been found in Helicidae and some other snails it appears to be completely absent from slugs (Razet and Dagobert 1968). In *Agriolimax reticulatus* less than 1·5% of the nitrogen excreted is in the form of gaseous ammonia (Garner 1970).

No data are available for the overall nitrogen balance in any mollusc. Nitrogen is lost by the slug in a variety of ways, of which probably the most important is excretion by way of the kidney, but some must be lost from the digestive gland by way of the faeces (the excretory cells have been shown to contain purines), and some from mucus secretion during locomotion and faeces production. Excretion of nitrogen in the form of gaseous nitrogen (Costa *et al.* 1968) or ammonia are also possibilities. The nitrogen excreted by the kidney has been followed in *Helix* and *Otala* (Table 6,), and the

Excretion

nitrogen lost as ammonia is estimated to be less than 5% of the purine nitrogen in active animals (Speeg and Campbell 1968a). From a determination of the rate of incorporation of glycine into the purine in the kidney it appears that the excretory potential of active snails is 78·2 mg purine/kg tissue/day.

It is believed that the majority of the purine reaches the kidney through the blood and estimates by Jezewska (1968) show that in *Helix pomatia* blood there are 2·6 mg uric acid, 1·6 mg xanthine and 1·2 mg guanine/100 ml blood. The value for uric acid exceeds the stated solubility of this compound in distilled water (Dawson et al. 1959), and as the solubility greatly increases with alkalinity it seems that the blood can probably carry this quantity only because of its alkaline pH.

It has been shown biochemically that molluscs synthesize uric acid in an identical manner to birds, insects and all other uric-acid synthesizing organisms (Lee and Campbell 1965). The digestive gland is the organ with the highest concentration of uric acid, apart from the kidney, and it is able to carry out the complete synthesis of this molecule. In many organisms guanine and xanthine are intermediates in the formation of uric acid but Speeg and Campbell (1968a) have shown, in *Otala,* that uric acid is synthesized at a faster rate than guanine and xanthine, which may indicate that uric acid is synthesized directly. Arginine biosynthesis has been studied in great detail in molluscs, because in vertebrates it is the origin of urea. Campbell and Speeg (1968a,b) have shown that the enzymes are distributed in all tissues and that radioactive precursors are rapidly incorporated into arginine which is then utilised extensively for protein synthesis. They conclude that the primitive function of arginine biosynthesis is protein synthesis not ureogenesis. It has now been shown that one of the enzymes—arginosuccinate synthetase—is absent from *Limax maximus* (Boonkoom and Horne 1968; Horne and Boonkoom, 1970).

The mechanism of kidney excretion has been studied in *Helix pomatia* and *Archachatina ventricosa* by Vorhwol (1961), and some of the details were subsequently verified by Martin, Stewart and Harrison (1965) for *Achatina fulica.* The fluid in the lumen of the kidney appears to be an ultrafiltrate of the blood and, as there seems to be little passage of material through the renopericardial opening, it is likely that it is derived from vessels in the kidney folds. In the primary ureter there is absorption of water and salts resulting in an increase in the pH. Further absorption of water occurs in the secondary ureter and some material is possibly secreted into the lumen by the ureter epithelium. The cytology of the ureters of *Helix* and *Lymnaea* where salt and water resorption have been clearly demonstrated is very similar to that found in *Agriolimax reticulatus*. Primary urine is formed at a rate of 8 ml/min/g in *Helix pomatia* while only 0·2 ml/g over a period of twenty days is actually passed to the outside as a semi-solid mass.

It is assumed by the various workers in this field that ultrafiltration must take place, if not through the walls of the blood vessels, then through the kidney epithelium. Ultrafiltration is the passage of fluid through a semipermeable membrane. For this to take place the hydrostatic pressure of the blood must exceed its osmotic pressure. In the pulmonate kidney there is no evidence for a semipermeable filtration area as in the vertebrate glomerulus; also the colloidal osmotic pressure of the blood varies from 6·6 to 9·9 atmospheres while the hydrostatic pressure in the haemocoel only reaches 0·03 atmospheres. It has been suggested by Boer (personal communication) that in *Lymnaea stagnalis* cytopempsis (transport of fluid across the cell within small vacuoles) may account for the passage of fluid into the kidney. It is however difficult to account for the lack of transport of haemocyanin if the mechanism involves cytopempsis. It is possible to speculate on the existence of another mechanism. Ultrafiltration is usually recognised by the passage of small molecules through a membrane leaving material of large molecular weight behind. In slug blood nearly all of the protein is haemocyanin, which has a molecular size of 300×335 Å. To remove this from the blood does not require a semipermeable membrane in the usual sense; a much coarser filter would be effective. The wall of the renal vessels appears to be a mass of collagen fibres equivalent to a depth filter; is it possible that this is sufficient to prevent the passage of haemocyanin? Filtered blood would then bathe the bases of the kidney cells: it is difficult to see how it could then pass into the lumen except by cell transport. If the kidney cells transfer the now haemocyanin-free blood by cytopempsis there will be an apparent ultrafiltration of the blood. Within the kidney cells the uric acid could be removed, since an acidification of the vacuole contents would be sufficient to precipitate the purines which are almost insoluble in acid solution.

5

REPRODUCTION, DEVELOPMENT, GROWTH AND GENETICS

REPRODUCTION

Although reproduction has been studied in detail in slugs there are still several functional aspects which are not understood. The most complete studies have been carried out on *Arion ater* (Smith 1966, Lusis 1961), *Agriolimax reticulatus* (Runham and Laryea 1968 and unpublished), *Philomycus carolinianus* (Kugler 1965), *Vaginulus borellianus* and *Laevicaulis alte* (Lanza and Quattrini 1964).

The hermaphrodite gland consists of many acini (sacs of cells) which open by way of efferent ducts into the hermaphrodite duct. Both efferent and hermaphrodite ducts are ciliated. Each acinus contains both male and female gametes, together with nutritive cells (Fig. 43). In very young animals the hermaphrodite gland is a simple sac filled with undifferentiated cells, but as it enlarges and becomes lobed, first oocytes and later spermatocytes differentiate. The young gonad continues to enlarge, becoming more and more lobed, until it assumes the character of the mature gonad. When sections of the acini are studied (Fig. 43) the oocytes are seen to be attached to the wall while the various spermatozoa stages fill the lumen. Evidence is accumulating that there is a continuous production of ova, sperm and nutritive cells in the mature gonad, except towards the end of the breeding season. After the cells which first filled the gonad have differentiated, further ova, sperm and nutritive cells arise from a ring of germinative cells at the neck of the acinus. The germ cells are continuous with the epithelium of the efferent ducts. As the cells migrate around the wall of the acinus the gametes gradually mature. Unwanted oocytes and probably also sperm are resorbed at the base of the acinus by nutritive cells. Some resorbing oocytes are usually visible in all acini and are easily recognised by the large vacuolar channels in their cytoplasm and the pycnotic nuclei. This sequence has been studied in most detail in the water snail *Lymnaea stagnalis* (Joosse, Boer and Cornelisse 1968) but studies on *Vaginulus* (Quattrini and Lanza 1965) and *Agriolimax* (Runham and Laryea 1968) lead to a similar conclusion. When the gonad is removed surgically from young animals, regeneration of the gonad takes place from the cut end of the hermaphrodite duct (Laviolette 1953). Following destruction of the gametes by radiation (Laviolette and Cuir 1959, and Joosse, Boer and Cornelisse 1968) or following resorption during starvation (Joosse, Boer and Cornelisse 1968) differentiation takes place from the germinative ring. The processes of spermatogenesis and oogenesis in slugs (Gatenby 1918, Quattrini and Lanza 1965, Richter 1935) are basically similar to those in other animals. Nothing is known however of the factors which determine the differentiation of ova, sperm, and nutritive cells from the apparently identical undifferentiated cells.

The first stage in the differentiation of the nutritive cells is a peculiar type of meiotic cell division (the highly coiled chromosomes form two groups and then the nucleus divides into two) termed spireme karyodiresis (Koshman and Serra 1967). As they migrate around the acinus they become larger and polyploid (Quattrini and Lanza 1965, and Serra and Koshman 1967). Some nutritive cells elongate and become covered with cells which differentiate into sperm. When the sperm have completed their differentiation they drop off into the lumen. The oocytes become surrounded by a small number of another type of nutritive cell forming a follicle. At ovulation the oocyte emerges from this follicle. The nutritive cells are characterised by the presence of very large numbers of granules and some multilamellate bodies in their cytoplasm. The fate of the nutritive cells after the gametes have been shed is unknown; but in animals which have completed reproduction the acinus becomes lined by columnar cells, and it is possible that these cells are transformed nutritive cells (Smith 1966). In those species which have a second breeding season gametogenesis restarts by the differentiation of these epithelial cells (Galangau 1964).

The rate at which gametes and nutritive cells differentiate is unknown. The number of gametes and the ratio between the numbers of sperm and ova can be altered by a variety of environmental factors. Low temperature delays development of the gonad (Lusis 1966, Smith 1966, Runham unpublished), while high temperatures lead to a reduction in the number, or even total loss, of the oocytes (Lusis 1966, Smith 1966). Continuous light of high intensity causes inhibition of gamete development (Pelluet and Henderson 1954) and low light intensities produce a more rapid maturation of the sperm. Complete darkness leads to gamete resorption so that the gonad resembles that of a senile animal (Pelluet and Henderson 1954, Smith 1966). Starvation retards the development of the gonad although sperm development is less affected (Richter 1935). A good diet promotes differentiation of the oocytes (Richter 1935). These conclusions have been reached by the examination of sections or by tedious cell counts (over 4,000 eggs may be present in a single gonad, together with seemingly infinite numbers of sperm).

The maturation of sperm is fairly rapid, so that although oocytes differentiate first they reach maturity after the sperm. When the sperm are released they are apparently immobile and probably are carried to the hermaphrodite duct by the cilia of the ducts. As sperm are lost from the hermaphrodite gland the acini shrink and appear empty. In *Arion ater* there tends to be a clear separation of the male and female phases of reproduction, so that most of the sperm are lost before the oocytes become mature (protandry). In *Agriolimax reticulatus* this separation is far less clear and judging from the histology, some eggs are laid before the major phase of spermatogenesis is over. The oocytes are not released from their follicles until just before egg laying. As the egg is much larger than the efferent duct it may be forced through by contraction of the muscle fibres in the acinar wall or it may pass through by amoeboid movement (Quattrini and Lanza 1965).

The hermaphrodite duct functions as a seminal vesicle, so that in mature slugs it is packed with sperm, appearing swollen and white in colour. During its storage in this duct some sperm, perhaps abnormal or aged, are resorbed by the lining epithelium.

At the time of egg laying the albumen gland is very large and the cells are completely filled with secretion. This secretion contains galactogen (a polymer of galactose), which appears to be the main energy source for the embryo, together with proteins, glycoprotein, calcium and perhaps other minerals (Bayne 1966 and 1967). The secretion is apparently released by the dissolution of the cell. At egg laying, each zygote receives some of this nutritive albumen. How this is controlled is unknown.

The structure of the common duct is extremely complicated. It consists of two highly glandular tubes which are in open continuity

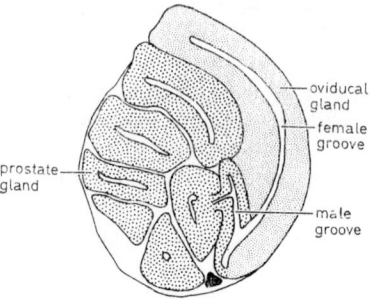

Fig. 34 *Agriolimax reticulatus*, transverse section of the common duct.

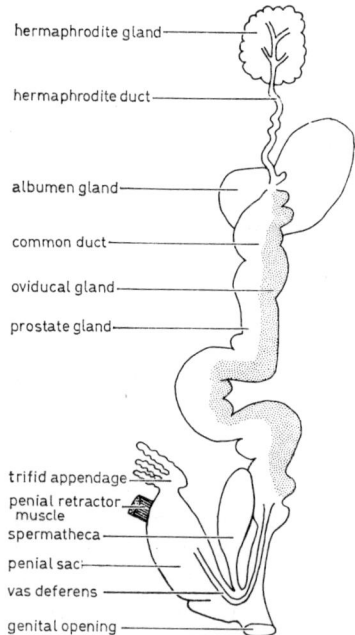

Fig. 35 *Agriolimax reticulatus* reproductive system (after Bayne 1966).

Reproduction, development, growth and genetics

along their length (Fig. 34). One of the tubes forms the oviducal gland which reaches maximal development immediately prior to egg-laying: the cavity of this tube is termed the female groove. The other glandular tube develops into the prostate gland which reaches maximum development at the time of copulation; the cavity of this tube is termed the male groove. At the lower end of the common duct the *vas deferens* and oviduct arise from the prostate and oviducal glands respectively. Both the vas deferens and the free oviduct join the large bulbous penis sac (Fig. 35). In the Veronicellidae the male and female ducts are separate (Fig. 36). Most slugs do not have a true penis: the genital apertures are merely opposed at copulation. During courtship in *Agriolimax reticulatus* the penis sac is everted, apparently by blood pressure. On the wall of this sac in *Agriolimax reticulatus* there is a large erectile sarcobelum and an evertile trifid appendage. The penis sac is extremely complex, having well developed

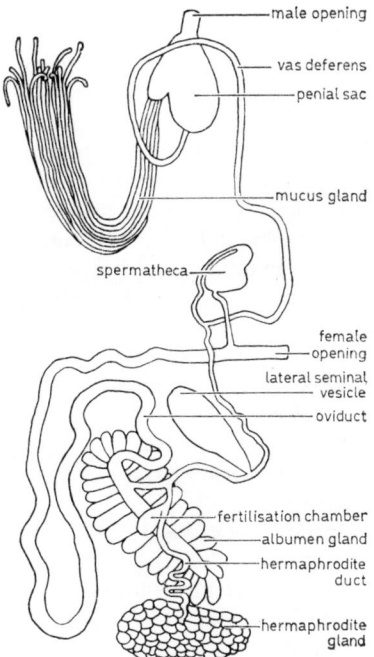

Fig. 36 *Vaginulus borellianus* reproductive system (after Lanza and Quattrini 1964).

ciliary tracts and several types of glands, but in no species has this structure been described in detail.

The spermatheca is connected to the penis sac by a short duct. In the mature animal the spermatheca after copulation is found to contain parts of the spermatophore or sperm mass but sperm are also found at other times. Eggs in various stages of breakdown are often present in the spermatheca after egg laying. It is usually assumed that sperm transferred at copulation are stored in this organ, but it seems likely that its main function is the disposal of excess genital products.

Courtship and copulation

Elaborate courtship sequences have been described for many slugs. In other species courtship has never been seen—perhaps because it occurs underground. In *Arion ater* (Gerhardt 1940, Quick 1947) the first sign of courtship is one animal pursuing another, often eating its mucus trail. Barr (1928) noted that in captivity one slug inspected the caudal glands of several other slugs until it found one that had a large globule of mucus present, which it consumed, and then started courting. In this animal the caudal gland only reaches maximal development with sexual maturity, so it appears that it may function as an indicator of the degree of sexual maturity in a potential mate.

In the second stage the pair start to circle around each other producing masses of mucus. On wet autumn mornings many of these circular patches of mucus may be visible, indicating extensive nocturnal courting. The genital atrium everts and can be seen as a large bluish swelling on the right side of the animal's body. They move closer together and the atria become closely applied to each other. Externally little can be seen except a line at the junction of the everted masses. Movement now stops and the slugs remain in this position for up to two hours. When they separate a spermatophore can occasionally be seen protruding from the spermathecal duct opening but it quickly disappears from sight as the atria are rapidly withdrawn.

In *Agriolimax reticulatus* (Gerhardt 1935, Webb 1961) courtship again occurs on the surface of the ground when it is damp, usually in the evening or at night. Courtship followed by copulation may take place several times in one night (Newell 1968). After chasing and circling phases of variable duration the sarcobelum is everted and as the animals get closer together it is used to stroke the body of the partner (Fig. 41) who may apparently bite it. Following varying

Reproduction, development, growth and genetics

periods of stroking and biting the animals stop moving and the remainder of the penis sac is everted, usually very quickly. The everted areas come into contact and the trifid appendages may cover the surface of the masses. The sperm mass is rapidly pushed out and passes almost directly to the slit-like opening of the spermathecal duct (Bailey personal communication) but some may remain stuck to the outside of the everted organs. Quite rapidly the penis sacs are withdrawn and the animals separate. The total time taken for this process may be up to ninety minutes, but the circling and copulation sequence usually takes less than thirty minutes. While this behaviour has been described several times, some details, particularly concerning sperm transfer, remain obscure.

The most spectacular courtships are undoubtedly those of the limacid species (Fig. 37). The most photographed has been the aerial

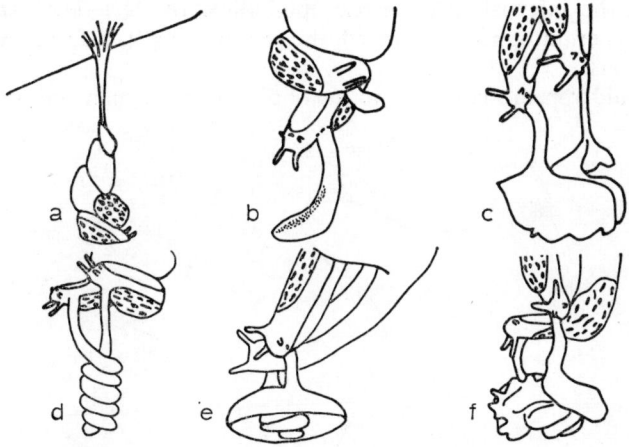

Fig. 37 Stages in the courtship of *Limax maximus* (after Chace 1953).

courtship of *Limax maximus* (Gerhardt 1934, Chace 1953). Pairing starts on a branch of a tree or at the top of a wall. For thirty to ninety minutes the couple follow each other in a tight circle caressing with their tentacles. Large amounts of mucus are secreted, and this forms a mucus string as the animals lower themselves from their support, at the same time entwining their bodies. The mucus string may extend to 45 cm. The penis sacs are everted. At first these sacs are club-shaped but then a terminal fan forms. The two penial masses intertwine forming a tight spiral; the upper coils of the spiral then expand to form an umbrella shape which eventually becomes lobed. Transfer of the sperm mass takes place while the penial masses are so extended (they may reach 10cm in length). The animals separate after sperm transfer, one usually ascending the mucus string, the other consuming it or descending to the ground. In *Limax redii* (Peyer 1954) the everted penis sacs are 70cm in length and the courtship takes from seven hours to more than a day.

During courtship a large number of sperm pass from the hermaphrodite duct into the male groove, where the sperm are coated with secretions from the prostate gland. In many slugs this secretion is hardened within a swelling of the vas deferens, the epiphallus, to form a spermatophore (Smith 1966). This spermatophore may have a very complex shape (Fig. 38), characteristic of the species, which mirrors the internal form of the epiphallus. In *Agriolimax* species no true spermatophore is formed, the sperm being transferred within a jelly mass.

It would appear that one function of courtship is synchronisation of

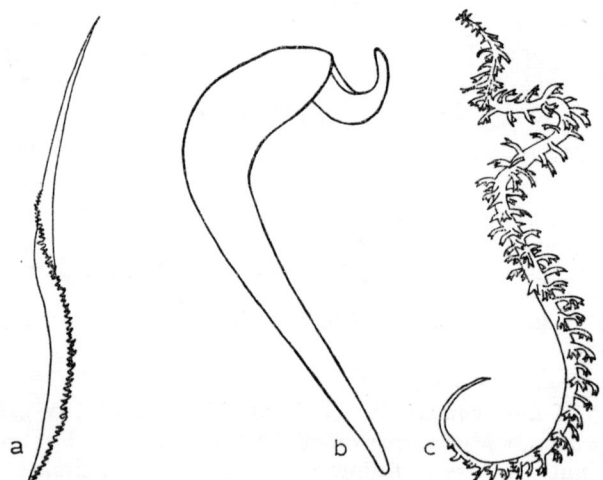

Fig. 38 Spermatophores.
   a. *Arion ater rufus* (after Quick 1960).
   b. *Arion fasciatus* (after Quick 1960).
   c. *Milax budapestensis* (after Newman unpublished).

the physiology of the slugs, i.e., the spermatophores are formed during this phase, the penis sacs or genital atria have to be everted, and the various duct openings are accurately aligned. As there is a characteristic sequence of events in courtship it seems likely that detailed behavioural analysis, like that which has been carried out on the courtships of other animals (e.g., sticklebacks), could be attempted here. Langlois (1963) has shown in *Limax maximus* that one of the pair (which he terms 'male-behaving individual') starts the chase and is the one that finally eats the mucus string, while the other ('female-behaving individual') always leads its partner up the tree or wall and is the first to leave after sperm transfer. It is known that in some species (e.g., *Arion ater*) transfer of sperm is not always mutual (p. 97); one partner may behave as a male, the other as a female and these roles are changed during subsequent matings (Williamson 1959).

Fertilisation

When the spermatophore or jelly mass is transferred at copulation it passes to the spermatheca where the outer material is digested away and the sperm released. The fate of these sperm is still uncertain but some undoubtedly pass to the hermaphrodite duct and fertilise the eggs. In many species of slug self-fertilisation is common, e.g., if the slug is isolated before sexual maturity it will still lay fertile eggs. *Agriolimax agrestis, A. meridionalis,* and *A. laevis* have been reared for twelve generations reproducing only by self-fertilisation (Maury and Reygrobellet 1963). A few species of slug (e.g., *Agriolimax reticulatus*) only rarely produce self-fertilised eggs.

The sequence of events leading from copulation to fertilisation of the ova has been studied in detail only for the pond snail *Lymnaea stagnalis* but the results agree with what has been published on other pulmonates including slugs (Horstmann 1955). After the sperm mass has reached the spermatheca in *Lymnaea* some sperm, about 1,000, pass up the common duct, while the majority remain within the spermatheca. The sperm are transported up the common duct by the cilia of the female groove; any remaining in the common duct after four to five hours are returned to the spermatheca by peristalsis. Mated snails ovulate much earlier than unmated ones and the stimulus was shown to be derived from the sperm mass within the spermatheca. Thus, if the spermathecal duct was cut before mating, or if the partner had been previously sterilised by cutting the vas deferens, no stimulation of ovulation occurred. Cutting the spermathecal duct after copulation had no effect. For the first two days after mating the eggs were shown to be self-fertilised but thereafter as many as 90% were cross-fertilised. This preferential cross-fertilisation appears to be due to the activation of foreign sperm by contact with common duct secretions. Sperm remaining within the spermatheca is broken down by secretions from its wall and resorbed.

The actual site of fertilisation in pulmonates is probably the junction of the hermaphrodite duct and albumen gland, where there is usually a small pocket, termed the fertilisation pocket. In some cases eggs as advanced as the blastula stage have been found within the gonad, implying that fertilisation occurred there. Wherever fertilisation does occur there is still the problem of explaining how the ova or alternatively the foreign sperm pass through the great mass of the animal's own sperm which packs the hermaphrodite duct. There is usually a short period between copulation and egg laying; eight to ten days in *Agriolimax reticulatus*; several weeks in *Arion ater*. During this time it seems likely that the sperm are stored in the upper part of the common duct, rather than in the spermatheca.

Egg laying

After ovulation the ova are fertilised and each zygote receives a layer of nutritional albumen (also termed perivitelline fluid) from the albumen gland. It then passes down the female groove. Within the oviducal gland the albumen-coated zygotes receive successive layers of material which form first the jelly layer, then, as they reach the lower part of the gland, the egg-shell layers (Fig. 39). Where the jelly and albumen layers come into contact a perivitelline membrane forms (Bayne 1966). The jelly layer consists of mucopolysaccharide (containing glucosamine and galactose) together with some calcium, and the shell layer contains a polysaccharide (a polymer of fucose and

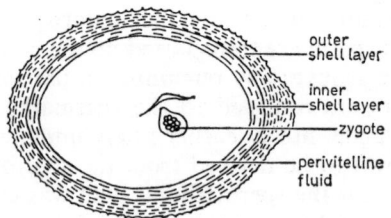

Fig. 39 *Agriolimax reticulatus* egg (after Carrick 1938).

galactose) protein and calcium (Bayne 1966). While the eggs of *Agriolimax* are completely transparent those of *Arion ater* have an outer layer of calcium carbonate crystals (secreted from the free oviduct) together with a lipid material secreted by a gland around the external genital opening (Smith 1966).

Many more ova are shed than are laid as eggs. Thus if *Vaginulus* is examined during egg laying it is found that while approximately 400 oocytes are shed only 25–225 are actually laid. The excess ova, some of which may be abnormal, are resorbed (Lanza and Quattrini 1964).

Characteristically, reproduction and egg laying occurs at a specific time of the year, but abnormal environmental conditions can alter the timing. Thus *Arion ater* normally reproduces in the autumn but maturation was retarded following an extremely dry summer so that the majority of the population did not lay their eggs until the following spring (Laviolette 1950a).

In the laboratory Segal (1959) has attempted to alter the period of egg laying by rearing slugs *(Limax flavus)* under various regimes of temperature and light. Egg laying still occurred at the normal time, so he suggested that it is controlled by an internal clock.

Eggs are usually laid in holes or crevices in the ground. In some cases the holes are first enlarged by the slugs, while a few species actively make a nest, as does *Vaginulus* (Colvin 1962) or *Laevicaulis* (Lanza and Quattrini 1964). Species vary in the number of eggs laid, their shape, size and the number of batches. *Vaginulus borellianus* lays 610–1,365 eggs in eight to thirteen batches and the eggs are in continuous strings resembling a necklace (Lanza and Quattrini 1964). *Agriolimax reticulatus* lays about 500 eggs in batches containing up to thirty-three completely separate eggs (Carrick 1938). In this species the maximum number of eggs is laid on soil 75% saturated with water (Arias and Crowell 1963). No eggs are laid when the soil is only 10% saturated and those laid at 100% and 25% fail to hatch. The drier the soil the deeper the eggs are laid (Carrick 1942). When the eggs are first laid they are very susceptible to extremes of temperature but as development proceeds their resistance increases considerably (they can then be stored for long periods in a refrigerator). Hatching takes eleven to twenty-one days at 20°C but often quite appreciable numbers of eggs are infertile.

When eggs are laid in moist soil they are in direct contact with the film of moisture over the soil particles. As soon as they are removed from contact with water they desiccate at a rate dependent on the humidity. There appear to be no specialisations of the egg which prevent water loss, but the larger the egg, and the greater the number of eggs in the group, the slower water is lost (Bayne 1968). Embryos are found to survive if after a loss of water from the egg (up to 80% by weight) they are immediately rehydrated, but longer-term and greater losses are fatal (Bayne 1968). At the other extreme, complete immersion in water for four days has little effect on development (Arias and Crowell 1963). Longer periods of immersion however considerably reduce hatching. The behaviour of the slug in selecting the site for laying eggs is thus extremely important for their survival.

DEVELOPMENT

Eggs are laid at various stages of development, some as advanced as the gastrula stage. In those species with a completely transparent egg (e.g., *Agriolimax reticulatus*) development is very easily followed (Carrick 1938). It has recently been found that the embryos of pulmonates will continue their development for several days when removed from the egg if they are kept in the albumen (Brisson 1968), so that it should be possible to study those species with opaque egg shells as well.

The zygote cleaves by typical spiral cleavage (Raven 1958) to form a blastula. Each cell's fate can be followed and predicted throughout development, i.e., the egg is determinate. The larval kidneys which have typical flame cells (Brandenburg 1966) and also accumulate crystalline excreta, develop soon after the blastula has invaginated. As the endodermal cells lining the gastrula absorb albumen they swell and form a large hepatic lobe (Fig. 40a). Soon after this the posterior part of the body becomes visible and a posterior sac on the tip of it rapidly enlarges (Fig. 40b). The hepatic lobe and the posterior sac contract alternately, circulating haemolymph around the body. In this they are aided by a contractile sinus termed the larval heart. The posterior sac is also the main respiratory organ and in some species becomes extremely large. Albumen is continually taken in through the blastopore, now the mouth, and is digested by the cells lining the hepatic lobe—this is the embryo's digestive organ. As the embryo increases in size the foot and mantle arise as double rudiments which later fuse. An ectodermal inpushing gives rise to the shell gland, the lower epithelium of which secretes crystals of calcium carbonate. The shell of pulmonate snails arises in a similar way and may become quite extensive before becoming external (Brisson 1968).

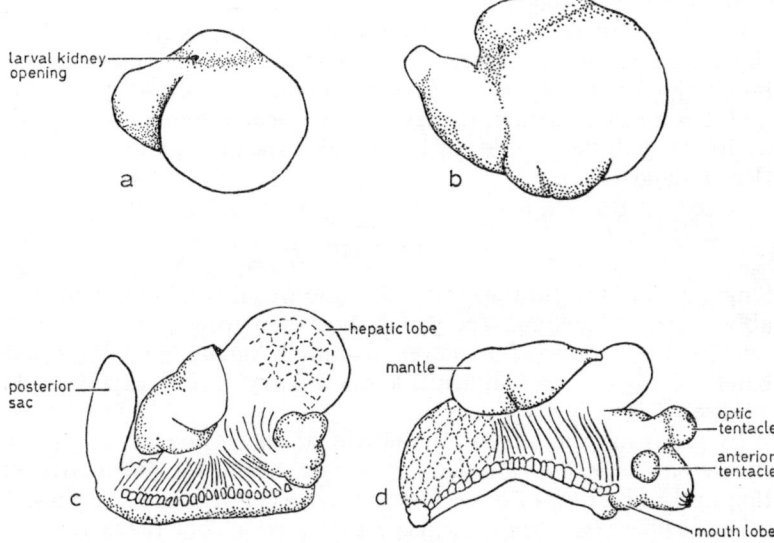

Fig. 40 *Agriolimax reticulatus*, stages in development (after Carrick 1938).

At about this stage torsion of the body through 180° occurs by differential growth, but the details of this process are not clear in these animals.

The posterior sac and hepatic lobe are gradually resorbed into the body as the heart and lung take over their functions (Fig. 40d). The digestive lining of the hepatic lobe becomes part of the digestive gland. Before hatching, the remaining albumen is ingested: in Arionidae it is stored within the digestive gland cells; while in the Limacidae it is stored in the lumen of the gland (Weiss 1968). The slug emerges from the egg after rasping the shell with its radula and pushing with its head.

The heart, pericardium and kidney arise from the mesoderm and the nervous system from the ectoderm. An ectodermal inpushing gives rise to the procerebrum, and the tube to the outside persists in the adult as the cerebral gland (Mol 1967).

The remainder of the cerebral ganglia and the other ganglia arise by epidermal proliferation. Eyes and statocysts arise by ectodermal inpushings. The digestive system is endodermal with the ectoderm contributing to the anterior parts of the system. Considerable controversy surrounds the origins of the reproductive system (Martoja 1964b): some authors claimed a double origin and others a single origin. The lung arises by an inpushing and may not be homologous with the pallial cavity of other molluscs. It is possible that it is a new structure, the pallial cavity being represented by the cavity beneath the mantle edge into which the anus and ureter open (Regondaud 1964).

GROWTH

Slugs hatch at various times after the eggs are laid and vary considerably in size at hatching. The young slugs then grow at very different rates. This variability, found even in animals hatching out of a single batch of eggs, causes difficulties when carrying out experimental work on slugs.

By following the weights of individual *Arion ater, A. subfuscus, A. intermedius, A. hortensis* and *A. fasciatus,* Abeloos (1944) discovered that in the first four of these species there were three growth phases: infantile phase (slope of growth curve 0·03) which had the fastest rate of growth; juvenile phase (slope of growth curve 0·01, less in *A. rufus*); mature phase (no growth). The discontinuities in the growth curves were termed pre-puberty and puberty. *A. fasciatus* showed only the juvenile and mature phases of growth. It was found that the maturation of the reproductive tract was closely related to the growth of the body (Laviolette 1950b, Abeloos 1944): during the infantile phase the gonad was undifferentiated; at pre-puberty differentiation of oocytes started; the juvenile phase was marked by extensive multiplication of the sperm-forming cells and an increase in size and complexity of the gonad; puberty was marked by the differentiation of the sperm; and the mature phase by growth of the reproductive tract. An essentially similar sequence has been derived from a study of reproductive maturation in *Agriolimax reticulatus* (Runham and Laryea 1968).

A phase of very variable length, termed the senile phase, follows the mature phase in *Arion ater* and *Agriolimax reticulatus* (Lusis 1961, Smith 1966, Runham and Laryea 1968); it is terminated by the death of the animal. In those species with a second reproductive season, similar changes in the gonad characterise the resting period which precedes renewed gamete differentiation (Galangau 1964).

Starvation and irradiation with both gamma and X-radiation stop growth. The duration of this growth inhibition following irradiation

Fig. 41 Copulating *Agriolimax reticulatus* (photograph by T. Byford).

Fig. 42 *Limax maximus* viewed from beneath while crawling on glass.

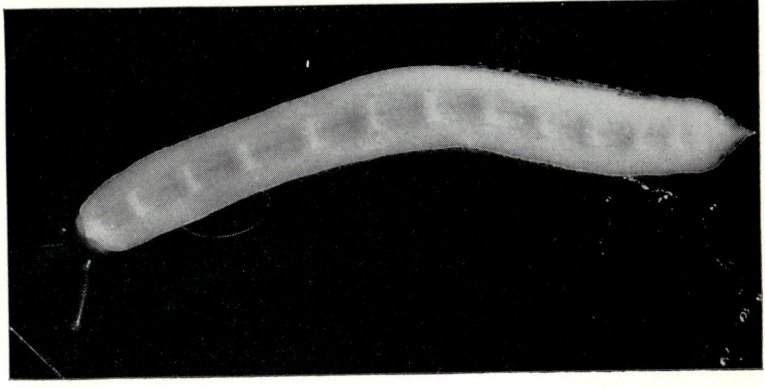

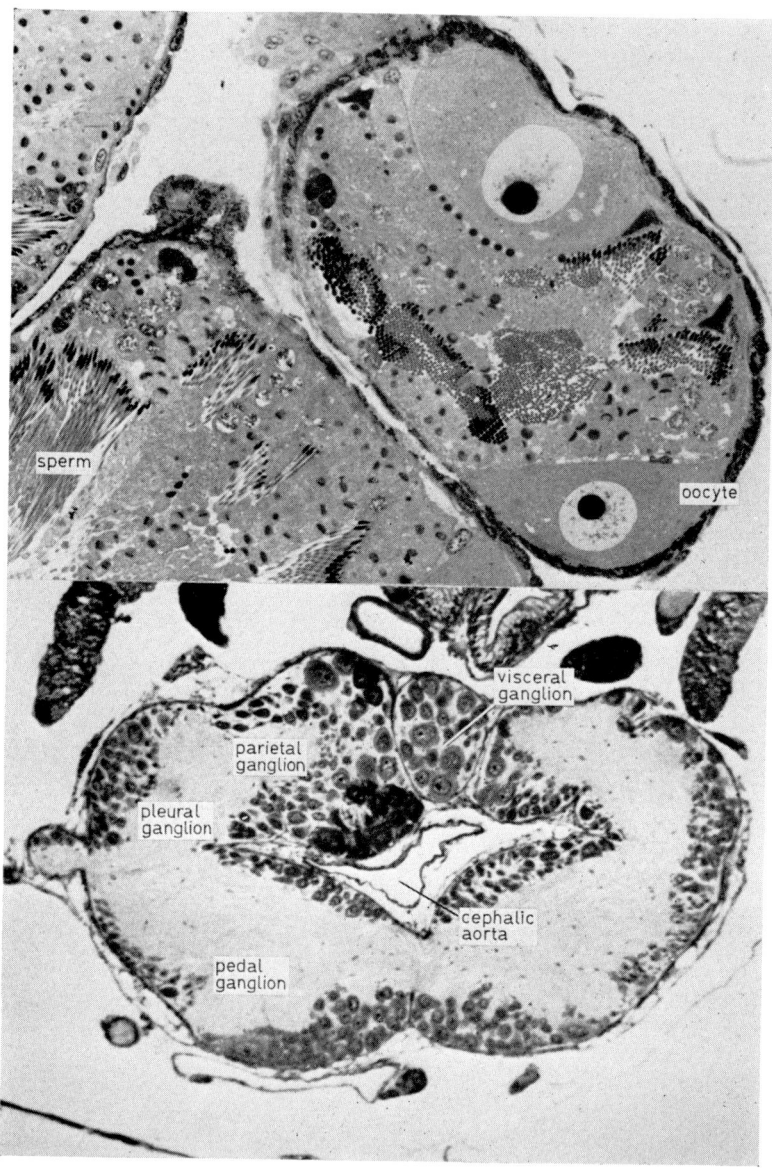

Fig. 43 (ABOVE) *Agriolimax reticulatus*, section of the gonad (1μ Araldite section stained toluidine blue).

Fig. 44 (BELOW) *Agriolimax reticulatus*, section through ventral part of the brain (stain azan).

(3,000–32,000 Röntgen) during the infantile phase is proportional to the amount of radiation received (Laviolette and Voulot 1961). Low doses of radiation (2,000 Röntgen) cause increased growth, apparently because the infantile phase is prolonged. Molluscs appear to be remarkably resistant to irradiation, for the doses used in this work were extremely high. The only result of irradiation that has been studied in detail is the effect on the gonad, where the initial loss of reproductive cells is followed by their regeneration.

GENETICS

Some aspects of the genetics of *Arion, Agriolimax* and *Philomycus* have been studied. As there are a large number of colour varieties of slugs (Taylor 1907) it is the inheritance of colour that has been most investigated. In *Philomycus bilineatus* (Ikeda 1937), *Agriolimax reticulatus* and *A. agrestis* (Luther 1915) there is a simple Mendelian inheritance of body pigmentation, pigmentation being dominant to albinism. Similar results have been obtained for three factors in *Arion ater ater* (Williamson 1959): black pigmentation (M) was dominant to brown (m); longitudinal banding (U) dominant to unbanded (u); and uniform pigmentation (F) dominant to streaking (f). The factors F and M, and U and M were found to be completely independent of each other but F was closely linked to U.

In addition to pigmentation differences there is also considerable variation in the morphology of the penial sac in *Agriolimax agrestis, A. laevis* and *A. meridionalis*. In self-fertilising animals the shape was inherited in every detail through twelve to thirteen generations (Maury and Reygrobellet 1963). The results of crossing the strains was not studied.

Using genetic markers it was found in the laboratory that in both *Arion* and *Philomycus* (Ikeda 1937; Williamson 1959) eggs were commonly self-fertilised, even following copulation. In *Philomycus* the proportion of cross- and self-fertilised eggs varied with the time of year, length of isolation and the time since copulation.

6

LOCOMOTION, MUCUS, SENSORY STRUCTURES, NERVOUS SYSTEM, ENDOCRINOLOGY

LOCOMOTION

Slugs use three main methods of locomotion: crawling with the foot flat on the ground; crawling with part of the body, usually the front, raised clear of the ground; and descending from a height using a mucus string. All of these methods appear to utilise the same mechanism.

If a slug is allowed to crawl on a glass sheet, alternating light and dark bands are seen to move forwards along the foot (Fig. 42), and where the slug has crawled a slime trail is left. Both features of locomotion are essential for movement to take place. Whenever a slug is crawling the forward-moving bands are always observed, and when stationary they are absent. If the glass is tilted as soon as the slug has been placed on it, the slug slides down, whereas if it is left for a while the glass can be inverted without the slug falling off. This is due to the presence of two types of mucus; one, secreted by the unicellular mucus glands of the foot epithelium, is thin and watery; the other, originating from the pedal gland, is poured out at the front of the foot, and is very sticky and thick. By adding coloured particles to the mucus Barr (1926) was able to show that the watery foot mucus is distributed over the surface of the foot, mainly from the centre towards the edges of the foot by cilia, while the pedal gland mucus passes backwards over the foot. After she removed the pedal gland from *Milax* by cautery normal locomotion was found to be impossible. The slug first raised its head to apply the mouth of the pedal gland to the substrate, as it normally does in this species when starting to crawl, then the back part of the foot moved up to the front the body thus forming a loop. Most animals then fell over, but some managed to progress by a looping movement. During this abnormal locomotion, waves were passing as usual forwards along the foot.

The dark and light waves have been studied in detail (Bonse 1934, Lissman 1945a,b) for *Helix* and several other molluscs. The dark bands are areas of the foot raised from the ground and they are narrower than the light bands which are closely applied to the ground. At the anterior edge of the light band there is a contraction of the surface which raises the foot from the substrate, while at the anterior edge of the dark band there is an expansion of the surface and application to the substrate (Fig. 43). Forward movement occurs when the foot is raised, not when it is applied to the substrate. Movement, however, depends on the adhesion of the light bands to the mucus trail.

Movement in animals lacking skeletons is usually achieved by means of a hydraulic mechanism: i.e., a fluid-filled space will shorten and widen by the contraction of surrounding longitudinal muscles, or lengthen and narrow by the contraction of surrounding circular muscles. In the limpet *Patella vulgata,* and apparently also in pulmonates (Jones 1968) there is a thick layer of dorso-ventrally oriented muscles in the foot with a thin layer of longitudinal and transverse muscle overlying it. Longitudinal and transverse cuts in the surface of the foot had little effect on locomotion. Jones suggests that contraction of the dorso-ventrally oriented muscles acting on blood-filled spaces causes lengthening of the foot (Fig. 45). In pulmonates Jones suggests that contraction of the dorso-ventral muscles results in a pushing forward of the front of the foot, while at the rear elongation is prevented by contraction of the longitudinal muscles.

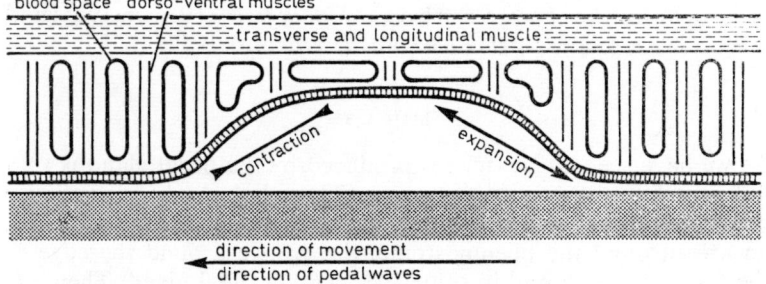

Fig. 45 Mechanism of pulmonate locomotion (after Jones 1968, Lissman 1945).

In *Limax maximus* the velocity of locomotion is directly proportional to the frequency of the pedal waves even when a vertically crawling slug is slowed by hanging weights from the mantle (Crozier and Federighi 1925b). When heavy weights are used locomotion ceases and the slug attaches itself to the substrate with the lateral margins of the foot which are not normally used in locomotion.

For movement to take place the foot surface must be raised during the passage of the dark waves from the sticky pedal mucus. It is possible that one of the functions of the watery foot mucus is to separate the foot from the pedal mucus. Machin (1964) has shown that mucus production depends on the hydrostatic pressure of the blood, as mucus is, in part, an ultrafiltrate of the blood (Burton 1965b). Contraction of the foot muscles increasing the hydrostatic pressure in the blood spaces will result in mucus production, and their relaxation will result in a reduction or stopping of mucus production. The cilia on the foot surface will move this watery mucus to the edge of the foot before the next phase of attachment.

As movement is produced by activity over the whole surface of the foot, if one region is raised the animal will still move forwards. Thus when the animal raises its anterior end this does not stop movement. When the animal is descending from a mucus string it can be seen that the mucus is produced from the pedal gland and that the locomotory waves are still present. In some species copulation takes place in mid-air at the end of one of these mucus strings (p. 89). In *Arion* the thick mucus produced by the caudal gland sometimes contributes to this mucus string (Barr 1928). The lateral movements of the body are produced by contractions of muscles in the body wall (Crozier and Federighi 1925a).

The speed of locomotion in slugs is not well documented; a speed of 17·6 cm/min has been recorded for *Limax flavus* (Brown 1955) and 0·79–2·88 cm/min for *Arion hortensis* (Verdcourt 1947).

MUCUS

As we have seen above, mucus produced by the pedal gland is very different from the foot-sole mucus. The whole body surface also produces mucus: several regions can be distinguished—the mantle, an area around the pneumostome, the groove around the edge of the foot, the head, and in some species the caudal gland. There are several types of mucus cell; some are unicellular epithelial cells while others are larger uni- or multi-cellular sub-epidermal glands which open through the epithelium. The cells vary in their histochemical properties, their appearance in the electron microscope and their distribution in the various body regions. In *Arion ater* four cell types have been described in the pedal epithelium (Chétail and Binot 1967) and three types in the general body surface (Wondrak 1968, 1969). Much work remains to determine the distribution of the cell types and to analyse the nature of their secretions.

The analyses of mucus that have been published refer to mixtures of all the different types of secretion. Burton (1965b) found in *Helix pomatia* that the solids in the mucus amounted to 120–350 g/kg water (slightly more concentrated than the blood). The variations in inorganic ion ratios suggest a dual origin for the mucus, one part being an ultrafiltrate of the blood and the other a secretion of the mucus cells. This suggestion is supported by the fact that mucus secretion is dependent on the hydrostatic pressure of the blood (Machin 1964). Mucus secretion, when repeatedly stimulated, had little immediate effect on the concentration of ions in the blood, but later there was a decrease denoting the uptake of ions for the formation of new mucus.

The organic components of the mucus are proteins together with a variety of complex carbohydrates. In the snail *Oxychilus alliarius* (Lloyd 1969) the mucus consisted of 6·8% solids and the solids were 8% inorganic, 77% protein and 15% carbohydrate. Analysis of the carbohydrate revealed the presence of fucose, glucose and galactose while in *Helix pomatia* (Fantin and Bolognari 1964) glucosamine and galactosamine have also been detected. The viscosity of the mucus is highly dependent on the divalent ion content, and it is interesting that the characteristic white and very thick mucus produced by *Agriolimax reticulatus* on irritation is rich in calcium salts. Some mucus is pigmented, e.g., the yellow mucus of *Arion subfuscus*; and pigment-containing mucus cells are recorded in the skin of *Arion ater* (Williamson 1959).

The pedal gland of *Arion*, *Limax* and *Milax* is either embedded in the foot-muscle *(Limax)* or lies in the haemocoel on top of the foot *(Milax)*. Along the sides of the large duct of this gland are large mucocytes of two types (Barr 1926, 1928, Chétail and Binot 1967) which secrete the pedal mucus. In *Milax* the duct extends posteriorly from the gland and contains two types of unidentified material, said to be excretory, one of which is crystalline. The duct of the pedal gland is ciliated, but in *Limax* where the gland is embedded in the foot it is possible that the foot muscles assist the expulsion of mucus.

When *Milax* starts crawling the front of the pedal gland is first applied to the substrate to stick the mucus, then as the animal crawls over it the mucus is drawn out by its tenacity.

SENSORY STRUCTURES

At the tip of each tentacle there is a large digitate ganglion with the cylindrical retractor muscle attached around its edges (Fig. 46). Other muscles are present in the wall of the tentacle at its base and these serve to orientate the tentacle. In the optic tentacles an eye is present to one side of the digitate ganglion.

The surfaces of the tentacles are kept moist by the numerous unicellular mucus cells in the epithelium. Large complex mucus glands open on to the surface of the mouth lobes.

The optic tentacles are sensitive to light and smell stimuli; the anterior tentacles to smell and perhaps taste; and the mouth lobes to mechanical stimuli (they can distinguish between different degrees of surface roughness) and perhaps limited taste sensitivity. When the tentacles are cut off they rapidly regenerate. In *Agriolimax agrestis* regeneration is complete in ten to twenty days and *Arion ater* thirty to sixty days (Chétail 1963). The regenerated tentacle in *Vaginulus borellianus* always lacks the eye (Renzoni 1969).

The structure of the eye of *Agriolimax reticulatus* resembles that of *Helix* (Newell and Newell 1968) which has been studied in very great detail (e.g., Schwalbach, Lickfield and Hahn 1963). There are however some differences, the most important of which is that an accessory retina is present in *A. reticulatus*. As slugs are normally active during the night it is not surprising that the eye is considered unlikely to be capable of detailed image perception but to be instead very sensitive to light intensity. There appear to be less than a hundred light-sensitive cells present (4,000 in *Helix*) each being surrounded by many pigment cells. At the upper surface of the light sensitive cells there are large numbers of microvilli (Fig. 46), and by analogy with other light receptors these are likely to be the light-sensitive areas. Contained within the cytoplasm there are very large numbers of small vacuoles, which are produced by the golgi bodies, and it has been shown that they accumulate vitamin A. It is therefore possible that these vacuoles represent the site of synthesis of a rhodopsin-like pigment (Eakin and Brandenburger 1968). Only one visual pigment is present in the snail *Otala,* absorbing maximally at 490 mμ. The lens originates during development by coalescence of granules produced by all the

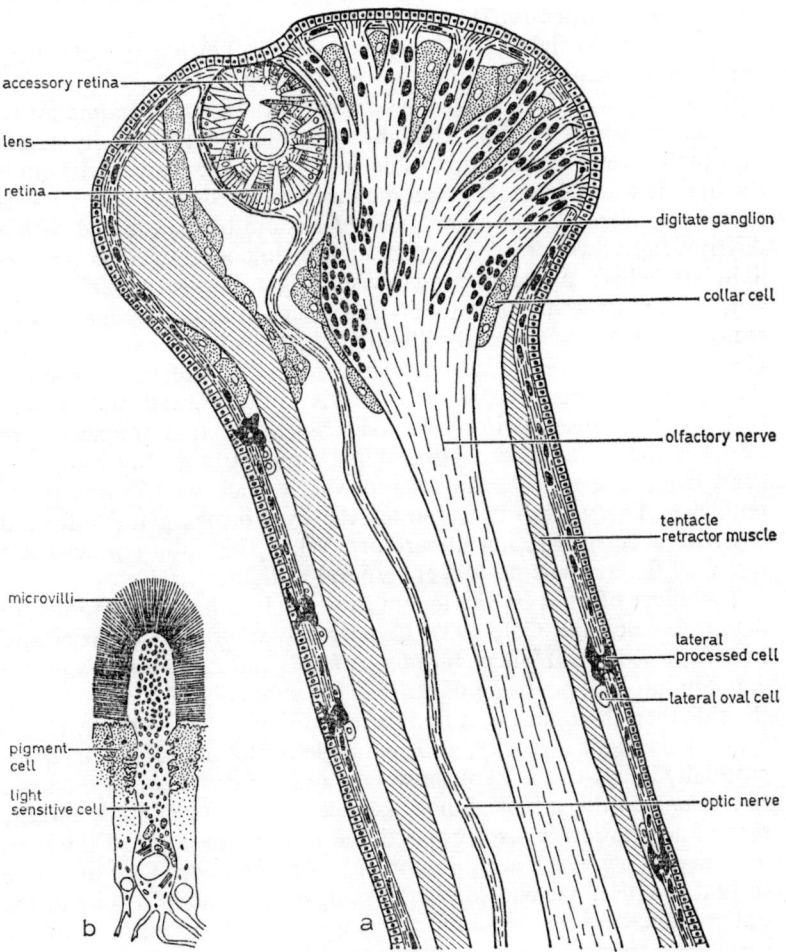

Fig. 46 *Agriolimax reticulatus*.
 a. Structure of tentacle (after Newell and Newell 1968, Chétail 1963, Röhlich and Bierbauer 1966).
 b. Retinal cells (Newell and Newell 1968).

cell types. The lens grows by accretion at its surface and in the adult, lens material (the vitreous) is secreted by the pigment cells (Eakin and Brandenburger 1967).

Slugs react to light in a complex manner. During the day they prefer shade (negative phototaxis), and if given a choice they are very active in the light but when they enter the dark they become much less active (Lewis 1969a). At night when slugs are normally active they prefer weak illumination to darkness (positive phototaxis) but if the light is too intense they become inactive. In the laboratory it was found that normal activity in *Arion ater* could be maintained with a daytime light level of 3,580 lux and night-time level of 5 lux (moonlight is 0·54 lux). While reacting to visible light slugs seem particularly sensitive to infra-red, so that if given a choice in daylight between two sectors, one being normally illuminated, the other covered with a glass dish containing water (which absorbs 50% of the infra-red), they preferred the covered area (Lewis 1967). Newell and Newell (1968) have suggested in *Agriolimax reticulatus* that the accessory retina is the infra-red receptor. When crawling, the slug moves its head from side to side and at intervals withdraws the tips of its tentacles. This withdrawal rotates the eye, bringing the infra-red sensitive accessory retina uppermost. When the optic tentacles are removed the animals no longer respond to light.

The effect of light on the locomotion of slugs has been studied in detail (Crozier and Cole 1929, Crozier and Wolf 1928, Crozier and Federighi 1925a). If when *Agriolimax laevis* and *Limax maximus* are crawling up a slope in the dark, they are suddenly illuminated from the side then their path is at first deflected in proportion to the logarithm of the light intensity, and the angle of illumination. The slugs gradually adapt to the light, however, and the deflection is reduced. In *Agriolimax* this adaptation took place within four minutes and the rate of adaptation is also proportional to the logarithm of the light intensity. When slugs are illuminated from both sides they orientate so that the photic excitation is the same on both sides. If one of the optic tentacles is removed and the slugs are illuminated from above they crawl round in circles, always crawling to the eyeless side. This effect is strongly influenced by temperature so that above 15°C the amount of crawling (degrees turned/cm of track) is directly proportional to the temperature. Below 15°C the amount of circling is determined by the rate of crawling which is inversely proportional to the temperature. The reactions of *Limax maximus* to light cease when they have eaten boiled potatoes: raw potato does not have this effect (Crozier and Libby 1925). Feeding sugars or injecting glucose produces a similar absence of phototaxis.

The structure of the digitate ganglion has not been well studied. It consists of a mass of neuropil with a layer of ganglionic cells to the outside. At its border with the epithelium it splits up into many fine processes which terminate in large numbers of sensory cells in the epithelium (Renzoni 1968). It has been shown many times that slugs react to smell. *Arion ater* and *Limax cinereoniger* (Kittel 1956) were tested for their ability to detect the presence of various fungi by placing them in the middle of a field at known distances from the fungi. The slugs reacted to smell stimuli in a characteristic manner; they would first crawl in the direction that they were set down, then they stopped, the tentacles moved from side to side and the slug next moved towards the food, not in a straight line but by a rather spiral path. Their ability to detect food depended on the species of fungus; thus *Arion* could detect the stinkhorn *(Phallus impudicus)* at 120cm; perhaps further. Of the species of fungi tested only those that could be detected by the slug (21 spp) were devoured. When food was moved to another place before the slug reached it, the slug stopped, retracted and extended its tentacles repeatedly then moved off in the new direction. On reaching food the slugs withdraw their optic tentacles while the anterior tentacles were kept applied to the surface of the food. If the optic tentacles were removed food could be detected only at much shorter distances (i.e. stinkhorn at only 20cm). When both pairs of tentacles were removed food could sometimes be detected, but only if placed directly in front of the mouth. When the food was covered with pine needles, after the animals deprived of tentacles had explored the surface of the food with the mouth lobes, they did not eat it. Similar results were obtained by streaking raspberry juice on filter paper and determining the ability of *Arion ater* to follow the streak. Animals with their tentacles removed wandered at random eating filter paper even without the raspberry juice. Kittel concludes that the optic tentacles are distance receptors for smell while the anterior tentacles are short distance smell and taste receptors. Taste receptors have also been found on the mouth lobes, the lips and the anterior edge of the foot of *Helix* (Kieckebusch 1953).

Similar experiments have been carried out with *Agriolimax reticulatus* (Griffiths unpublished). When the attractiveness of a variety of foods was measured, cut potato was detected over the longest distance (cut potato>carrot>lettuce>germinating wheat> mouse muscle>*Ranunculus repens* leaves>undamaged potato).

Removal of the small tentacles decreased the slugs' ability to detect potato but optic tentacle removal had a more profound effect (Fig. 47). Some ability to smell remained after removal of both pairs of tentacles. These experiments were performed on a sheltered lawn at night and the slugs responded only in the absence of any wind.

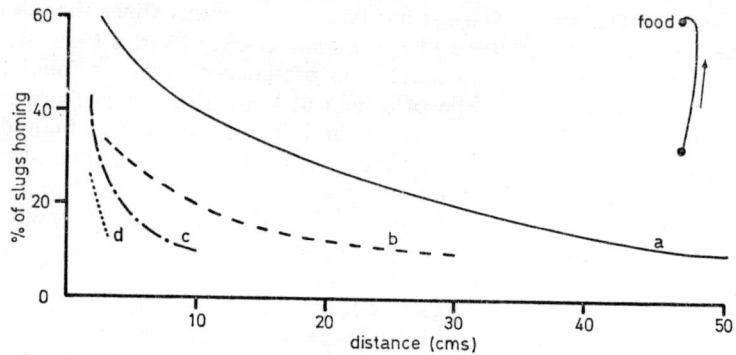

Fig. 47 *Agriolimax reticulatus*. Effect of tentacle removal on the ability to smell fresh potato (from Griffiths unpublished).
a. Normal animals.
b. Anterior tentacles removed.
c. Optic tentacles removed.
d. Both pairs of tentacles removed.
Insert—path taken by slug when homing on the food.

Statocysts

These are situated on the pedal ganglia but are innervated by nerves from the cerebral ganglia. They are paired hollow structures containing many small calcareous bodies, the statoliths. In *Limax maximus*, *L. flavus* and *Arion ater* (Wolff 1969) the statocysts are lined by an epithelium consisting of thirteen large flat ciliated cells, termed hair cells, interspersed with small columnar cells. These large cells appear to be the sensory cells as they are associated with large axons. In *Helix* the hair cells secrete the statoliths, but they are secreted by the small columnar cells in regenerating *Lymnaea* statocysts (Laverack 1968e, Geuze 1968).

Statocysts are normally associated with gravity perception. There has until recently been considerable controversy as to whether it is the muscle tensions on the two sides of the body, or the statocysts, which gives rise to the clear geotactic response of pulmonates. Operative experiments have now clearly shown that the statocysts at least play the dominant role (Geuze 1968 and Coleman 1970). If slugs are placed on a slope they usually crawl up it at an angle to the perpendicular (Wolff 1927). In *Agriolimax laevis,* if the slope is less than 45° then orientation is not clear, but above this the steeper the slope the more perpendicularly the slugs crawl. Angles greater than 90°, which means that the slug is partially inverted, cause a similar response. Very similar behaviour is shown by the snails *Lymnaea* and *Helix* but when both their statocysts are removed they wander aimlessly. In air, when crawling up a slope, the normal snail's path deviates to the right; but when the right statocyst is removed they deviate further to the right, when the left is removed they deviate to the left, of vertical. Electrophysiological studies on *Arion ater* show that when the slug is rotated around its longitudinal or transverse axis there is a change in the frequency of impulses in the statocyst nerve. The statocysts can detect both rotation and orientation. Not all of the nerve fibres in the statocyst nerve were sensory; a few efferent fibres also innervate the statocyst.

Temperature sensitivity
Dainton (1954a) concluded that temperature is the main factor controlling the activity of slugs. Changes of temperature of as little as 0·1°C/hr below 20°C stimulated activity. Other workers have failed to obtain these results (Lewis 1969a). Short-term changes in locomotor activity in *Agriolimax* following alterations in temperature have been correlated with the nervous activity of the ganglia. Thus cooling produced a transient increase in both locomotor and ganglionic activity, while heating produced a corresponding decrease (Kerkut 1959).

Humidity sensitivity
With the use of humidity gradients it can be shown that slugs normally prefer more humid areas. At the boundary between a dry and a humid zone *Arion ater* execute head movements and then crawl into the damper area (Lewis 1969a). When the difference between the two zones is less than 30% R.H. *Milax gagates* will accumulate in the drier zone during the day but will migrate to the damper one at night (Bonavita 1967). As Lewis has pointed out it is difficult to believe that humidity receptors could be present in the skin of an animal whose surface is always moist. Evaporation of water from the skin, the rate of which is proportional to the humidity, produces cooling.

It is therefore possible that it is not humidity but skin temperature changes that are detected. Loss of water causing an increase in the osmotic concentration of the blood has been shown to decrease nervous activity in the pedal ganglia of *Agriolimax*. Dilution of the blood which leads to increased locomotion also produces an increase in the nervous activity of these pedal ganglia (Kerkut 1959).

Wind sensitivity

Slugs show a clear response to wind currents (Kalmus 1942, Dainton 1954b). Very weak air currents result in an extension of the tentacles towards the source of the wind. If, however, a slug is crawling and strong currents impinge on the tail it crawls faster, while if from any other direction it turns and crawls away from the wind. As mentioned above (p. 73) the rate of water loss from snails is least when they crawl downwind.

Factors affecting the normal behaviour of slugs

During the day slugs are usually found inactive in sheltered situations, while during the night they are active. It has been shown that if *Arion ater* are kept in constant conditions they show a persistent diurnal rhythm of activity of approximately twenty-three hours, which after two weeks becomes less clearly defined (Lewis 1969b). Animals that had lost their rhythm only regained it when subjected to a light-dark regime approximating to twenty-four hours. Lewis suggests that a rhythm of less than twenty-four hours enables the animal to adjust to slow changes in day length. Thus observations on *Arion ater* showed that a burst of activity occurred in the early evening: the slug emerged from its resting place, and if it was light the animal withdrew and became inactive again for a short time. Should it be dark earlier than usual, because of cloud or heavy rain for example, then the slugs are active earlier than usual. At sunrise they sought shelter and then became inactive. There therefore appears to be an internal rhythm which is entrained by the normal diurnal rhythm of light and dark.

When slugs become active they show irregular periods of eating, resting, locomotion and mating (Newell 1968). What stimulates these activities and determines their duration is unknown. When exposed to an air current they crawl downwind at a fast rate but when they reach shelter they crawl more slowly and may become inactive.

Marked *Agriolimax reticulatus* released into a strange environment crawl large distances (10–20 m), but when they arrive at a suitable

Locomotion, mucus, sensory structures, nervous system, endocrinology 109

sheltered habitat they tend to remain within that area for many days (Moens, François and Riga 1966). It thus seems likely that shelter is of prime importance for determining the distribution of slugs.

THE NERVOUS SYSTEM

The nerve ganglia and nerves of *Agriolimax reticulatus* are shown in Figs. 48 and 49. In most of the organs, but particularly in the foot, digestive and reproductive systems there are complicated plexuses. These can be studied by staining techniques and also by physiological methods. When organs are removed from the body they continue their rhythmic functions for some time, and as these rhythms are changed by treatment with drugs affecting nervous tissue it seems likely that the plexuses have an important controlling function (Minker and Koltai 1961).

The structure of the nerve ganglia (Laryea 1970) is extremely complex (Fig. 44). Around the outside of the ganglion there is a thick sheath containing much collagen, muscle fibres, and various connective tissue cells, some of which contain material which appears to be secretory. To the outside of the ganglionic tissue there are many nerve cell bodies, while the centre of the ganglion is filled with a mass of nerve fibres (neuropil). Interspersed with the nervous elements there are large numbers of supporting cells (glial cells) which are apparently nutritive. Some glial cells contain masses of glycogen, others (sheath cells) have thin sheet-like processes which are wrapped around the nerve cells. The nerve cell bodies vary very much in size. There are a few giant cells (400 μ) whose nuclei are highly polyploid; containing more than 10,000 times more DNA than other cells (Kuhlmann 1969). At the other extreme there are groups of very small cells (10 μ) in the procerebrum. Within the cells there are many vacuoles with very variable staining properties, and their size varies from 200 to 4,000 Å (Gerschenfield 1963). Vacuoles containing pigmented material are present in the cytoplasm, particularly in the giant cells—the function of this material is unknown. Processes from the sheath cells push into the nerve cell bodies forming a very extensive network, the troposphongium. It is assumed that this increases the surface area for metabolic exchange. In *Ariolimax* (Turner 1966) it can be concluded that there is a constant arrangement of at least the giant and medium-sized cells.

The nerve cells in the brain may have one or several axons and dendrites. These processes are surrounded by large numbers of glial

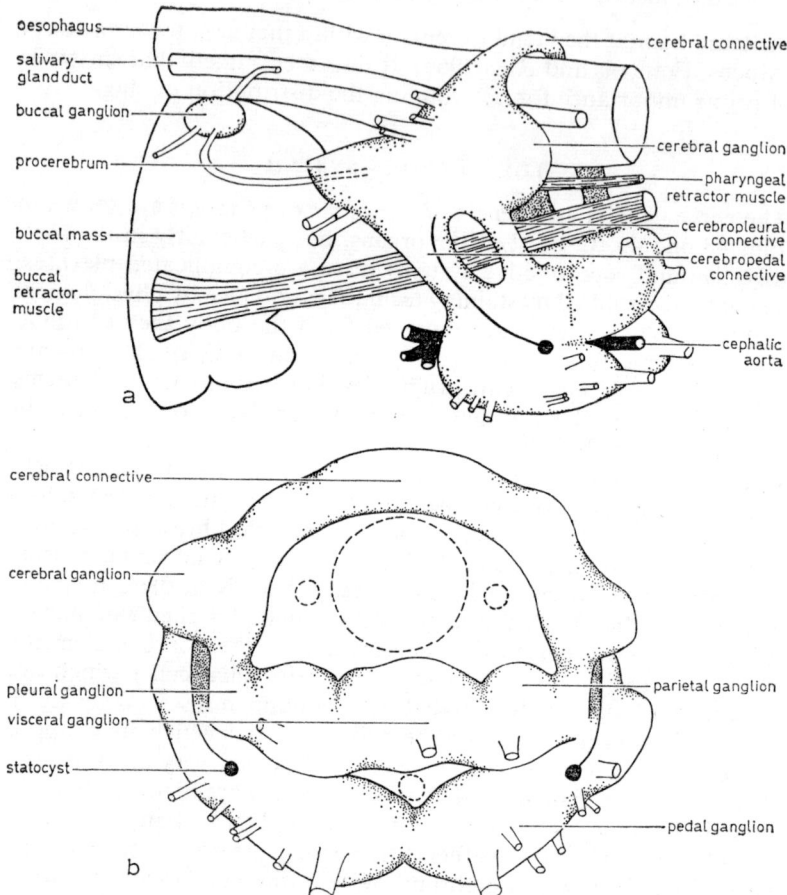

Fig. 48 *Agriolimax reticulatus* brain (after Laryea 1970).
    a. Lateral view.
    b. Posterior view.

cells. The nerve axons within the nerves are similarly surrounded by supporting and nutritive cells, termed Schwann cells. The nerves usually run in close association with arteries, often within the same sheath. The synapse (junction between two nerve cells) is simple i.e., there is simply a close apposition of the cell membranes. In pulmonates it has appeared until recently that all synapses were axo-axonic

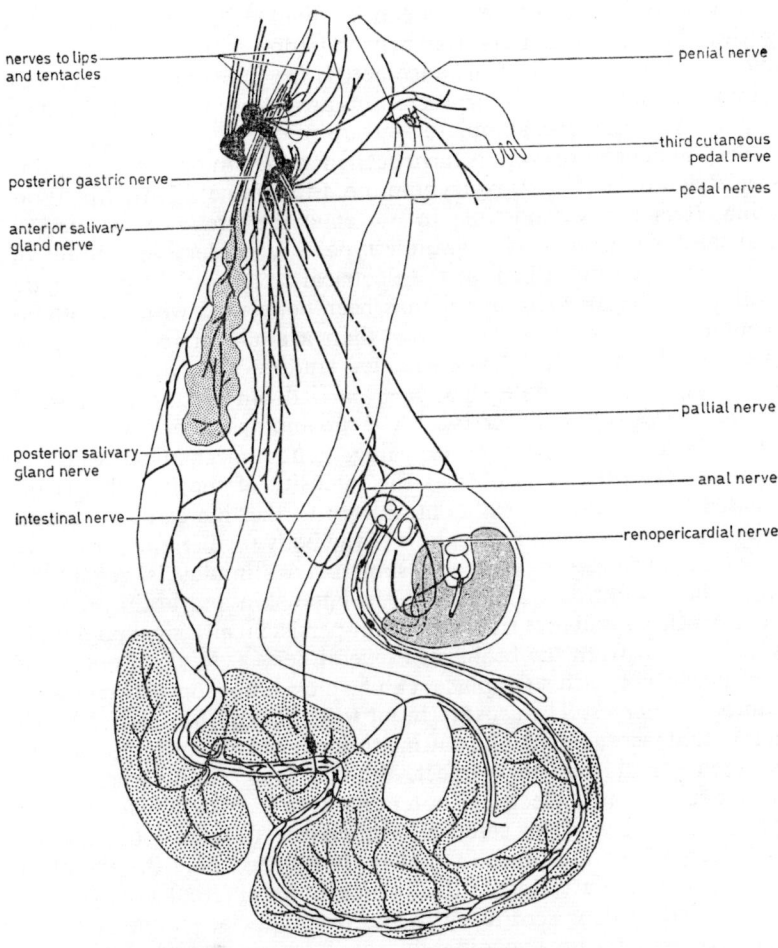

Fig. 49 *Agriolimax reticulatus* nervous system (after Laryea 1970, Walker 1969, Garner 1970).

(only between axons) but recently axo-somatic synapses (between axons and nerve cell bodies) have been described in the procerebrum of *Limax* (Zs-Nagy and Sakharov 1969). Nerve cell bodies are quite frequently found along the nerves, in places forming small ganglia.

The secretions of nerve cells can be divided into two main categories. Neurotransmitters (or neurohumors) are released at the synapse, travel the short distance across the intervening space, and stimulate another cell. Neurosecretions are released into the blood or connective tissue and stimulate a group of cells at a distance.

Several neurotransmitters have been identified in pulmonate brains—acetylcholine, 5-hydroxytryptamine (dopamine), 5-hydroxytyramine. It is not yet possible to correlate the presence of a given transmitter with a specific granule type. A considerable amount of work has been published on the electrical activity of the pulmonate brain, but the majority of this has been concerned with the fundamental properties of nerve rather than a study of its function in the animal. One of the few correlated studies of nerve activity and behaviour in pulmonates has been carried out on the pond snail *Lymnaea stagnalis* (Vlieger 1968) and the findings seem applicable to slugs. When the animal is crawling, gentle mechanical stimulus induces very little response; as the intensity of the stimulus is increased there is first a local contraction; further increases of stimuli result in the contraction of regions of the body further away from the point of stimulus. If a strong short-lived stimulus is given, the whole body contracts violently. In isolated skin and brain preparations gentle stimuli give rise to fast non-persistent electrical potentials which spread from the brain into several nerves. As the strength of stimulus is increased so impulses can be picked up from an increasing number of nerves. The sensory input to the brain thus spread to all nerves that were studied, and cutting the connectives and commissures between ganglia had little effect. When strong stimuli were given, slow but persistent electrical potentials were found in a few of the large nerve fibres. The cerebral commissure differed from all the other regions of the brain in that it transmitted only the impulses arising from these sudden strong stimuli. As different areas of the skin varied in their sensitivity to the two types of stimuli it seems likely that there are two different types of sensory cell concerned with the reception of mechanical stimuli.

ENDOCRINOLOGY

Discrete endocrine organs comparable to those of vertebrates are rare in molluscs and endocrine control appears to be brought about by secretions of the nervous system—neurosecretions. Much work has been carried out on molluscan neurosecretion but the earlier work based on light microscope studies is of doubtful validity. The neurosecretory 'stains' (paraldehyde fuchsin and chrome haematoxylin) in fact stain a wide variety of materials including neurotransmitter granules, lipid and glycogen (Gabe 1965, Simpson, Bern and Nishioka 1966). With the electron microscope many granules are visible in nerve cells, and it has been suggested that neurosecretion is associated with the large dense granules. Such granules have been shown in some cases to contain neurotransmitters, and also a wide variety of granules are present in cells that are clearly neurosecretory. For a positive identification of a neurosecretory cell it is essential to determine the site of release. A demonstration of seasonal or other changes in the amount of neurosecretion is also difficult to interpret, as it has to be established whether the presence of large amounts represents storage, or indicates that large amounts are being secreted. Similarly the presence of only small quantities of neurosecretion in the cell may mean that it all is being secreted. It is therefore essential in these studies to relate changes in cell contents to experimental studies on the function of the secretion. Such exhaustive studies have been carried out for some aspects of the physiology of the pond snail *Lymnaea stagnalis* (Lever and co-workers).

Several endocrine organs have been suggested in slugs: dorsal bodies, cerebral glands, Semper's organ, arterial gland, and the tentacles. Semper's organ appears to be a collection of mucus glands and sensory nerve ganglia associated with the mouth lobes (Laryea 1970), and no endocrine function has been proved. The cerebral glands represent the stalks of the ectodermal inpushings that gave rise to the procerebra, and they may contain stainable material. In *Arion ater* the cerebral glands are larger in young animals than in older ones, so that they could be associated with growth (Mol 1961). However, they are very small in other slugs. Dorsal bodies are groups of cells which extend over the cerebral ganglia and down on to the pleural ganglia (Laryea 1970). They are intimately associated with the connective tissue of the brain and discharge their secretions into the haemocoel. While the function of the dorsal bodies in slugs is not clear it has been suggested that they are concerned with osmotic control (Kuhlmann 1966), and in *Helix* injections of calcium or magnesium chloride cause an increase in their size (Nolte and Machemer-Röhnisch 1966). In *Lymnaea stagnalis*, however, extirpation experiments have shown that the dorsal body tissue controls vitellogenesis—the laying down of yolk in the egg (Joosse and Geraerts 1970). The arterial gland apparently secretes into the haemocoel but its function is at the present time completely unknown (Laryea 1969). When reproduction has been completed in *Agriolimax* the breakdown of this gland precedes the animal's death. Several neurohaemal areas have been described in slugs (Mol 1967; Laryea 1970) but their function is completely unknown. In *Lymnaea stagnalis* the neurosecretory cells in the dorsal part of the cerebral ganglia appear to control the growth of the animal and the maturation of the reproductive tract (Joosse and Geraerts 1970).

While the function of these suggested endocrine tissues is in the main obscure, there is a growing body of experimental data clearly demonstrating the endocrine control of reproduction.

Studies of animals castrated by infections with parasites indicate that there is a close relation between the gonad and the reproductive tract; experimental studies confirm this (Abeloos 1943 and Laviolette 1954). In *Limax maximus* castration is simply performed by removal of the rear portion of the animal (Abeloos 1943). Following this operation the albumen gland and common duct remain the same size at the time of operation, or regress. The penis sac is not affected by the operation. Laviolette (1954) confirmed and considerably extended these results using various Limacid and Arionid slugs. As the various species mature at different times of year an extensive series of transplantations of gonads and reproductive tracts could be performed. The conclusions from these experiments are that the maturation of the albumen gland and common duct are controlled in some way by the gonad. The controlling factor has to be blood-borne, as the transplants were free in the haemocoel, but no cytological evidence for any hormone-producing cells could be found in the gonad, nor did injection of gonad extracts have any effect. In *Agriolimax reticulatus* (Runham, Bailey and Laryea unpublished) when an undifferentiated tract (oviducal, prostate and albumen glands completely undeveloped) is put into the haemocoel of an animal laying eggs, the oviducal and albumen glands enlarge very considerably while the prostate gland remains undeveloped (this situation is never seen in normal animals). If an undifferentiated tract is put into the haemocoel of an animal in the early male phase of its development (i.e., the gonad is full of sperm and the ova are still very small) the prostate glands enlarge considerably while the oviducal and albumen glands remain small. When undifferentiated tracts are transplanted into animals at later male stages, the whole tract may enlarge. These results are interpreted as implying the presence of two blood-borne hormones; one (present early in maturation) controls the male gland, and the other (present later in development) controls the development of the female glands.

As the tract maturation hormones do not appear to originate from the gonad it would seem likely that the nervous system is involved. In *Ariolimax carolinianus* it is claimed that the hormone(s) is in fact produced by the cerebral ganglia (Gottfried *et al.* 1968).

The development of the gonad itself appears to be controlled by antagonistic hormones originating from the tentacles and cerebral ganglia. When the tentacles are removed there is an increase in the number of oocytes. Injection of tentacle homogenates produces an increase in spermatogenesis while injection of brain extracts produces an increase in oocyte production (Pelluet and Lane 1961, Pelluet 1964). It is suggested that the tentacles produce two hormones, one inhibiting the production of the brain hormone, and the other stimulating spermatogenesis, while the brain produces an oogenesis-stimulating hormone These results appear to have been confirmed for various snails, by the organ culture of brain and tentacles together with gonads (Guyard and Gomot, 1964). In *Vaginulus*, however, removal of the tentacles has no effect on the gonad nor on growth (Renzoni 1969).

In *Ariolimax columbianus* the production of tentacle spermatogenic hormone appears to be controlled by steroids released from the gonad (Gottfried *et al.* 1968). It has been claimed that the collar, lateral oval, and lateral processed cells (Fig. 46) in the tentacle are the source of the tentacle hormones (Lane 1962) but recent electron microscope studies show this to be unlikely (Röhlich and Bierbauer 1966; Bierbauer, Török and Teichmann 1965; Rogers 1969).

7

ECOLOGY

Having examined the physiology and the behaviour of slugs, mainly under artificial conditions, we must now study the living animals in their natural habitats. We have to see how the slugs react to a variety of weather conditions, how they avoid or succumb to their enemies (including man) and how they react to competition from other animals (including other slugs) competing for the same resources. It is this interaction between the animal (or a population of animals) and the environment which forms the basis of ecology.

The ecology of a partly subterranean, aggregated and fairly sparsely distributed animal like the slug is never easy to study. The usual purpose of making an ecological study of a particular species, or group of species, is to discover reasons for its distribution and fluctuations in numbers in a particular habitat. Very few such studies have been made on slugs and none for longer than three years. Only one study has been carried out on a site where *Agriolimax*, *Arion* and *Milax* species were all present in large enough numbers for definite conclusions to be drawn about their distribution and abundance. In the following account, this study of the ecology of *Agriolimax reticulatus*, *Arion hortensis* and *Milax budapestensis* in a market garden in Northumberland (Hunter 1966, 1968a,b,c) will be discussed in detail, and other studies referred to where necessary.

METHODS USED IN ECOLOGICAL STUDIES

Sampling methods

If the ecologist is to observe the rise and fall in numbers of a slug population, he must first have a method of estimating its density. It is obviously not possible to count the slugs directly, since they spend most of their time underground. It is therefore necessary to

Ecology

obtain a sample of slugs that is representative of the population as a whole. There are a number of methods for obtaining samples and these must be tested before deciding which one to use. An excellent review of sampling methods is given by Southwood (1966).

The usual method for soil invertebrates is to take a sample of soil and then extract the animals in the laboratory. As a general principle each sample must include a number of 'sampling units' taken at random from the area being studied. This reduces the risk of sampling only from dense or sparse parts of the population and thus drawing biased conclusions. The number and size of units within each sample depends on the distribution and numbers of the animals in the habitat. If the population is dense and evenly distributed, fewer sampling units will be required than if the population is sparse and aggregated.

In the market garden study, sampling units covering an area of less than 20–30 square inches contained very few or no slugs. To be

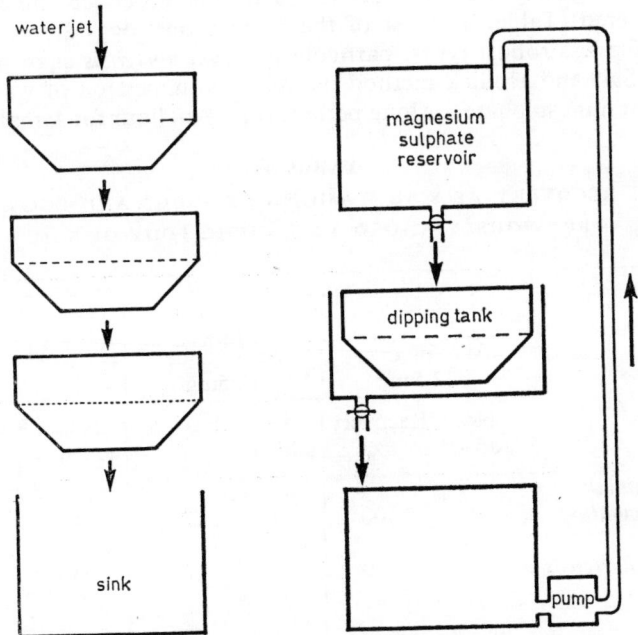

Fig. 50 Soil washing. The soil is washed through the three sieves and the sieves are then placed in the magnesium sulphate solution to float out the slugs.

certain that a reasonable proportion of units would contain more than one or two slugs a unit size of about one foot square was required. It was also found that many of the slugs lived up to a foot underground so the size of unit chosen for samples was a twelve-inch cube. Facilities were available for digging and extracting only twelve of these units at any one time.

The most commonly used method of separating animals from soil is the Salt and Hollick process (Salt and Hollick 1944). The soil is placed on a bank of three sieves (Fig. 50)—for slugs, three, ten and thirty meshes to the inch—and broken down with a powerful water jet. The large slugs are collected on the ten-mesh sieve and the small slugs and eggs are washed down to the fine mesh. The sieves, with their residues, are then agitated in a solution of magnesium sulphate or brine with a specific gravity of at least 1·17 mg/cc. All the organic matter rises to the surface of the solution so that slugs and eggs can be picked off. The main disadvantage of this method is that many of the very young slugs are destroyed by the water jet and are not recovered (Table 7). Most of the more robust eggs are recovered but the less robust types, particularly *Arion hortensis* eggs, are not. The Salt and Hollick method necessitates immersion in water and magnesium sulphate for long periods (up to an hour for samples with

TABLE 7

RECOVERY BY SOIL WASHING OF SLUGS AND EGGS, PREVIOUSLY ADDED TO 1 CUBIC FOOT OF SOIL

Species	Slugs				Eggs	
	Wt. of over 12·5 mg		Wt. of under 12·5 mg			
	No. added*	Recovery %	No. added*	Recovery %	No. added*	Recovery %
Agriolimax reticulatus	15	100	8	62·5	11	90·9
Arion hortensis	15	100	12	66·7	13	15·4
Milax budapestensis	15	100	13	85·6	15	100·0

* The operator did not know the number added until after the experiment.

a high proportion of vegetation) and slugs can lose up to one third of their body weight before examination; this is a major disadvantage for life cycle studies, where the weight of the slug is used as a measure of its age.

A much less laborious method of extracting slugs is slowly to flood the soil with water. South (1964) developed this method for his study of *Agriolimax reticulatus* on grassland. This species does not penetrate more than about 10 cm. deep on grassland and it was possible to take a complete turf, 30 cm sq, as the sampling unit. These units were brought into the laboratory and individually placed on end in plastic dustbins. Water was then added at regular intervals until the turfs were submerged in three to four days. The slugs crawled up the turf, just above the water level, and were picked off as they reached the upper edge.

Cultivated soils, however, crumble as they are dug from the field and, for the market garden study, the flooding extraction method was modified. Plastic washing-up bowls, fitted into plastic dustbins of the same circumference, were used for this (Fig. 51). The bowls had holes in the bottom so that soil placed in them could be flooded from below. The bins were filled with water up to the bottom of the bowls and the water level was raised half an inch every 12 hours to immerse

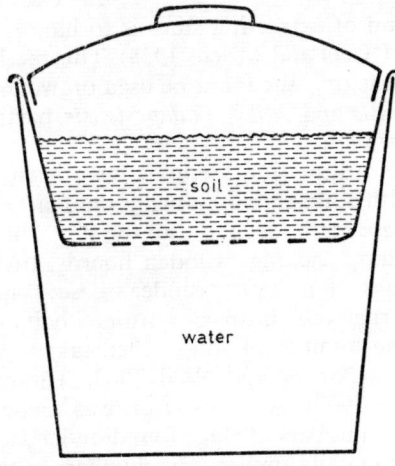

Fig. 51 Soil flooding. The soil is placed in a washing-up bowl with a perforated bottom, which is then placed in a plastic dustbin. When the water level is raised the slugs crawl to the top of the soil where they are picked off.

the soil in four to five days. The slugs were induced to move upwards as the water level rose and were picked off the soil surface or the lid of the bin. About 90% of the slugs in the soil were extracted by this method, but of course eggs were not (Table 8). The greatest

TABLE 8

RECOVERY BY FLOODING OF SLUGS PRESENT IN 16 CUBIC FEET OF SOIL

Species	Nos. Extracted	Not Extracted*	Nos. Damaged†	Success %
Agriolimax reticulatus	78	7	2	91·8
Arion hortensis	49	7	2	87·5
Milax budapestensis	49	6	0	89·1

* Found later by soil washing.
† Slugs damaged during digging of samples and not included in the calculation of % success.

advantage of this method over soil washing is that the extracted slugs are in good condition.

A further method of extracting slugs is to hand-sort the soil and pick out the slugs (Bruel and Moens 1958). This method can be used only where the soil is dry. It cannot be used on wet or sticky soils in which *Arion hortensis* and *Milax budapestensis* particularly become encased in soil particles.

Other population estimation methods have been used and these were compared with soil sampling during the market garden study. The simplest of these methods is to trap the slugs on the soil surface by laying down damp sacking, wooden boards, tiles or boxes and collecting any slugs that gather under these traps (Getz 1959, Howitt 1961). Alternatively, heaps of poisoned bait can be placed on the ground and the number of slugs killed taken as a measure of population density (Barnes and Weil 1942, Thomas 1944, 1948, Webley 1962, 1964, 1965). A comparison was made in the market garden between the numbers of slugs found under sacking traps and the numbers found in soil samples. The numbers of the three species of slugs under traps were not proportional to the numbers in the population as found by soil sampling (Table 9).

When the weather was dry or cold, fewer slugs were found under

TABLE 9
COMPARISON OF THE SPECIES COMPOSITION OF TRAP SAMPLES AND SOIL SAMPLES

Species	Traps		Soil Sample		Totals from Traps	
	No.	%	No.	%	Sunny	Cloudy
Agriolimax reticulatus	328	61·8	14	17·3	138	190
Arion hortensis	135	25·4	40	49·4	22	113
Milax budapestensis	68	12·8	27	33·3	23	45
Totals	531	100·0	81	100·0	183	348

traps than when it was warm or wet, even though there had been no change in the population on the plot. No slugs were found under traps when it was very dry or frosty since slugs are not active on the surface in unfavourable weather.

Another simple but unsatisfactory method of sampling slugs is to spend a certain amount of time (about half an hour) searching over a particular piece of ground for active slugs at night (Barnes and Weil 1944, 1945, Bett 1960). When this method was tested in the market garden, a much higher percentage of large *Agriolimax reticulatus* was found than was suggested from soil samples (Table 10). It seemed that

TABLE 10
COMPARISON OF THE SPECIES COMPOSITION OF NIGHT-SEARCHING SAMPLES AND SOIL SAMPLES

Species	Night-searching sample		Soil Sample	
	No.	%	No.	%
Agriolimax reticulatus	300	85·23	27	27·84
Arion hortensis	19	5·39	46	47·42
Milax budapestensis	33	9·38	24	24·74
Totals	352	100·00	97	100·00

although numbers of smaller and darker slugs were present many were not noticed. This method, like trapping, cannot be used in dry or frosty weather when the slugs are not active.

A simpler and even less accurate method of assessing slug numbers is to measure the feeding activity of the population. This can be done by placing wheat grains (Duthoit 1961, 1964) or potato tubers on the soil surface and noting the amount of slug damage that occurs. Slug damage to both wheat and potatoes is quite distinctive and is described in Chapter 8. From a comparison of Duthoit's method with soil sampling in the market garden there was some evidence that a large number of records of feeding activity, over a long period of time, could bear some relation to the numbers of slugs present. Activity however depends so much on temperature and humidity that this method would be unlikely to give information on changes in numbers from month to month (Table 11).

A more sophisticated sampling technique is called the 'marking-

TABLE 11

THE EFFECT OF TEMPERATURE AND HUMIDITY ON POPULATION NUMBERS ESTIMATED BY BAITING

Mean No. of slugs	Mean Temp. °C.	Mean No. hours high humidity	Mean No. grains damaged
98	8·74	25	5·7
131	12·63	17	8·3
188	13·00	35	22·3
178	15·33	27	3·0
217	13·77	45	14·7
273	13·05	43	36·0
274	10·74	45	54·7
224	7·91	57	46·7
115	1·50	40	1·0
107	3·31	46	3·3
110	4·50	32	4·7

release-recapture' method. For this it is necessary to collect a large number of slugs and mark them in some way, for example by simply feeding the slugs on agar jelly containing the dye neutral red. Another marking procedure is to feed the slugs on lettuce containing radio-active phosphorus (Newell 1965), but this requires expensive equipment to check which slugs have taken up the marker. The

Ecology

marked slugs are released on to the study area and allowed to mix freely with the resident population. A further sample is taken some time later, and the ratio of marked to unmarked slugs in the sample can be used to indicate how many unmarked slugs are in the study area. This method was used to estimate the population of *Agriolimax reticulatus, Arion hortensis* and *Milax budapestensis* in part of the market garden being studied. Three hundred *Agriolimax reticulatus,* three hundred *Arion hortensis* and one hundred and eighty *Milax budapestensis* were captured by night-searching, marked and released. The recapture sample, again collected by night-searching, contained 300 *Agriolimax reticulatus* (15 marked), 18 *Arion hortensis* (2 marked) and 33 *Milax budapestensis* (1 marked). This suggested that the numbers in the study area were approximately 4,000 *Agriolimax reticulatus* (compared to an estimate of 6,500 from a soil sample taken at the same time), 2,000 *Arion hortensis* (11,000 from the soil sample) and 3,000 *Milax budapestensis* (6,000 from the soil sample). It was therefore concluded that this method could only be accurate if a great deal of time was spent obtaining the recapture sample.

This work on sampling showed that although other methods may be useful for investigating slug activity the most economical method for obtaining estimates of slug population density was soil sampling. It was decided to take twelve sampling units (each of one cubic foot of soil) every four weeks. Each unit was sub-divided horizontally into four layers, 0–3 inches deep, 3–6 inches, 6–9 inches and 9–12 inches, so that the vertical distribution of the slugs could be studied. The slugs were extracted by soil washing during 1963, so that the pattern of egg laying could be studied, and by flooding from January 1964 to March 1965, when sampling ended.

RESULTS FROM FIELD STUDIES

The results from the study in the market garden illustrated two aspects of slug ecology. Information was provided on the distribution of these slugs in space and on the seasonal variation in numbers of slugs through time.

Distribution in space

There were significant differences between the vertical distribution of the three species and between the vertical distribution of all slugs at different times of year. For example, in 1963 most *Agriolimax reticulatus* (97%) were found in the top three inches of soil whereas only 81% of *Arion hortensis* and 72% of *Milax budapestensis* were found at this level (Table 12). Also, the top three inches of soil

TABLE 12

PERCENTAGE OF SLUGS AT DIFFERENT DEPTHS IN THE SOIL, 1963

in deep	Agriolimax reticulatus	Arion hortensis	Milax budapestensis
0–3	96·9	80·8	72·0
3–6	2·7	9·3	13·4
6–9	0·4	6·4	9·0
9–12	0·0	3·5	5·6

contained more than 60% of the slugs except in December, January and February, when it appeared that the slugs had been driven deeper by frost, and in June, when drought had a similar effect (Table 13). Similarly, the proportion of *Agriolimax reticulatus* eggs in the top three inches was greater than the proportion of *Arion hortensis* or *Milax budapestensis* eggs.

TABLE 13

PERCENTAGE OF SLUGS IN THE TOP THREE INCHES OF SOIL, 1963

	Agriolimax reticulatus	Arion hortensis	Milax budapestensis
January	94·4	32·1	19·4
February	84·2	35·7	16·7
March	100·0	86·3	58·3
April	100·0	95·0	59·3
May	100·0	95·9	89·5
June	80·0	7·1	61·5
July	88·5	74·5	75·0
August	100·0	79·7	79·5
September	96·6	97·0	96·7
October	98·7	100·0	100·0
November	100·0	100·0	100·0
December	100·0	98·4	100·0

Ecology

The horizontal distribution of slugs also showed significant variations on the study area. There were fewer *Milax budapestensis* (but not other species) on that part of the plot where a drain had been excavated the year before sampling began. It was assumed that most of the slugs had been killed when the drain was dug (see Chapter 8), and that *Milax budapestensis* was slower to recolonise this ground than the other two species. It was also found that the numbers of *Agriolimax reticulatus,* but not the other two species, were directly related to the dry weight of vegetation growing on the surface.

Large-scale surveys have been made of the horizontal distribution of slugs (Gould 1961, Bruel and Moens 1958). These surveys have shown that slugs are most common in heavy soils where the moisture is retained at high levels during the summer, in contrast to the lighter soils which become dry. Dense populations are also commonest where crops such as peas or beans have given shelter during the summer, and more important, few slugs are found where this cover does not occur. Surveys have also shown that there are definite differences in the species composition of slugs on different cropping systems. Where mainly grass or cereals are grown, *Agriolimax reticulatus* tends to be the commonest species; where root crops or potatoes are frequently grown, *Milax budapestensis* or *Arion hortensis* is commonest. *Agriolimax reticulatus* is able to move more quickly and further than the other two species, and so is better adapted to search for surface shelter. Thus when there are few soil spaces available as on grass or cereals, more of this species can survive. *Agriolimax reticulatus* has also a higher reproductive capacity and populations can withstand a high rate of depletion from drought or frost. *Milax budapestensis* and *Arion hortensis* survive drought and frost by living deeper in the soil and are consequently less likely to be killed by farm implements on crops which require frequent cultivation.

Some species of slugs, for example *Agriolimax laevis,* are able to survive only in marshland or damp habitats. Others, for example *Limax tenellus* and *Limax cinereoniger,* seem to prefer woodland. Still others prefer unusual habitats such as trees or stone walls *(Lehmannia marginata)* or drains *(Limax flavus).* Apart from casual observations on the prevalence of these species in various sites, there has been no comprehensive research on their distribution. No reasons for differences in species distribution have been given.

There are also, of course, major differences in the geographical distribution of the various species. Although many European forms

(*Agriolimax reticulatus, Arion ater, A. fasciatus,* for example) have been introduced into America, none of the American species has been introduced into Europe. There is evidence that the distribution of some species is still rapidly changing. *Agriolimax caruanae,* previously mainly confined in Britain to the west coast, has recently been reported as a pest in East Anglia. On a smaller scale, slugs have been reported as well established on land reclaimed from the sea only three years previously.

Aggregation and dispersal

A facet of the study of both horizontal and vertical distribution is the marked aggregation exhibited by slugs. Aggregation can be measured as 'the co-efficient of dispersion' (Clarke and Milne 1955, Milne 1962) and, when this was plotted against time for the market garden study, it was shown that the slugs were particularly aggregated immediately after breeding seasons. Clearly, the recently hatched young had not yet dispersed. Some aggregation is found throughout the year, presumably mostly due to the uneven distribution of cracks and spaces in the soil.

The ability of slugs to disperse into previously unpopulated areas was studied by South (1965) in the same market garden. *Agriolimax reticulatus* were marked by feeding them on an agar jelly containing neutral red. They were released from a point in the centre of a plot which had traps of brick, 20cm square, arranged in concentric zones around it. The number of marked slugs gathering under traps in each zone was recorded at daily intervals after release. The data from this experiment showed that the slugs dispersed for two to five days until there was an even distribution over a central part of the plot, but then the slugs moved out no further. They appeared to establish a 'home range' when they had become less crowded, and remained in this range until observations were concluded five days after release. It is not clear how this range is maintained. Newell (1966) made time-lapse photographic records of the nocturnal excursions of *Agriolimax reticulatus* and these slugs seemed to show definite tendencies to return to the place of shelter from which they had started. However, Newell's observations were made in a limited area (1·3 metre × 1 metre) with only six holes for shelter. Duval (in press), from records of the movement and sheltering places of slugs in a garden in Hertfordshire, reported that the movement of slugs tends to follow a circuitous pattern, bringing the slugs at the end of each excursion into a position near their starting point. Thus, although

Ecology

slugs will not deliberately return 'home' there is a tendency for them to find the same shelter on two or more consecutive occasions.

Further studies on the dispersal of slugs (Pinder 1969; Johnston unpublished) have demonstrated that young slugs disperse for a longer period before settling down to a 'home range'. Also, the speed of dispersal and length of time spent dispersing depend on the amount of shelter (greater dispersal with less shelter) and weather (greater dispersal when warm, moist weather encourages activity). It is not clear whether any of these experiments can be regarded as conclusive. Unless large numbers are taken in samples after release, it is very difficult to differentiate between the 'normal' curve of dispersing objects (for example, dispersing inanimate particles) and dispersing animals with a tendency to establish a 'home range'. This is particularly complicated with animals such as slugs whose dispersing activity may suddenly cease because of unfavourable weather.

Distribution in time

It was a little more difficult to evaluate the changes in numbers that occurred on the plot from month to month. The high degree of aggregation meant that it was possible for sampling units to be taken by chance only from areas where there was a 'clump' of slugs or from a relatively empty area between clumps. Thus, even fairly major changes in numbers could be accounted for by 'sampling error' and it was not until a definite trend began to appear that conclusions about changes in numbers from samples could be made with any confidence. The study in the market garden did not fully establish how numbers can be expected to fluctuate throughout the year but did provide a general picture of the seasonal fluctuations in numbers (Fig. 52). Although there are differences between species, most of the major decreases in numbers occurred between late November and mid-February. There was also an indication of a decline in June, although this was masked in *Milax budapestensis* by the May hatch of the new generation. It can therefore be concluded that numbers decline in the driest and coldest months. There is not sufficient evidence to come to any more definite conclusion. Normally a full-scale population study of an animal would require seven to ten years of regular sampling before the ecologist would be confident enough to say how much seasonal fluctuations in numbers can be explained by the various environmental factors. No one has yet done this kind of long-term study on any species of slug.

Many field studies of slugs must be interpreted with reserve

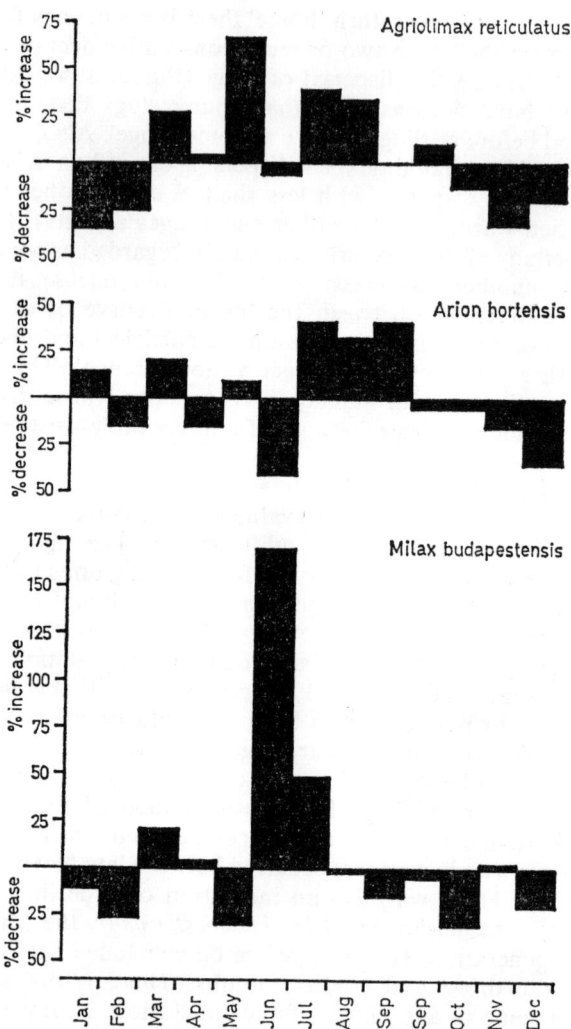

Fig. 52 Increases and decreases in the numbers of slugs. Samples were taken every 4 weeks.

because of the inadequate sampling methods used. For example, Getz (1959) studied *Arion fasciatus, Agriolimax reticulatus* and *A. laevis* in Michigan, USA. He used wooden box 'traps' as his sampling technique and showed that more *Arion fasciatus* were active after rain than *Agriolimax reticulatus* or *A. laevis*. All species entered 'hibernation' (i.e., they no longer appeared under the traps) at the same time in the autumn. It is not clear from this study, however, how much of the variation in numbers sampled is due to differences in activity and how much is due to changes in population density.

DISCUSSION OF THESE STUDIES

We can now discuss the results of sampling studies in a more general form. A complete investigation of the ecology of slugs would involve detailed studies of all factors operating within communities where they are found. This has been done in small, isolated communities of other animals, where it is possible to estimate the way in which energy is passed from the atmosphere to vegetation ('primary production'), and then to herbivorous animals ('primary consumers'), and to predators, parasites and scavengers ('secondary consumers'), and so on. This sort of study does not help us to understand the distribution of a single species, or how the numbers are regulated, unless there is a shortage of energy for that particular species. It is most unlikely that there is a shortage for slugs as they feed on a wide variety of food and can withstand starvation for long periods. If a shortage of energy does not affect the distribution and numbers of slugs in nature, what does? We have seen that drought and frost can reduce numbers, but is there any other factor in the environment of slugs which can affect their density?

The distribution and abundance of animals are determined by their intrinsic capacity to survive and increase in numbers, and by the power of external environmental factors to affect this capacity. There is widespread controversy among ecologists over the way in which environmental factors affect the capacity of populations to increase in numbers. Many theories of natural population control differ only because they consider different lengths of time (e.g., single point in time, or over many generations) or different sizes of population (a limited area like the market garden or a major geographical area). It is necessary, even when giving a fairly simple account of the ecology of a single group of animals, to specify these. In this chapter, we will consider only factors as they affect limited populations (i.e., single fields or gardens), at any particular point in time. Factors will be described in terms of their effect on the proportion of the population killed, not in terms of absolute numbers killed (or surviving), or the probability that an individual will survive.

The controversy is beginning to resolve itself as some of the older theories are discarded. It is now apparent that environmental factors affect populations in two ways. Weather has a direct effect on a population and the intensity of this effect is in no way determined by the density of that population. Thus where there is no competition for shelter, a severe frost will kill the same proportion of a population however dense or sparse that population is. Factors whose effect is not related to the density of the population are termed 'density-independent'.

Some factors vary in intensity with the density of the population. Thus the effect of parasites on a population, for example, is not only dependent on the direct influence of some external factors, but also on the availability of their food (the density of the host population). If a host population is sparse, the parasite population will have more effect than if it is dense. Thus, this factor can be called at least partly 'density-dependent'. There is only one 'perfect density-dependent factor': the competition for resources in a population (Milne 1957, 1962).

Ecologists are still not agreed on the relative importance of the effect of density-dependent and density-independent factors on natural populations. It is impossible to conceive a general theory without some element of density-dependent control. If all populations were able to increase until some chance factor depleted them, many species would overcrowd themselves to extinction. Some ecologists infer that density-dependent factors alone are operative in determining the density of all populations at all times (Nicholson 1954, 1957, 1958); others claim that they are operative most of the time (Richards and Southwood 1968); while some ecologists claim that density-dependent factors operate only rarely (Milne 1962). The present writers hold the latter view. The very great majority of species are extremely rare (Wilson 1968); evidence is only available to show that a tiny proportion of species become so dense that they compete for resources. Most species increase gradually in numbers until some environmental factor (usually weather in a regular, seasonal pattern) depletes them. The weather never releases (using the concept of Morris 1963) their intrinsic ability to increase in numbers for long enough for them to achieve dense populations (Andrewartha and Birch 1954).

A further source of misunderstanding among ecologists lies in the failure of many workers to take into account the whole 'life system' of the population (Clarke *et al.* 1967), i.e., the population itself must be considered as part of the environment. The capacity of a population to increase in numbers is the basic parameter of population studies and this has been called the 'innate capacity for increase'. This term has been defined as 'the maximum rate of increase attained at any particular combination of temperature, moisture, quality of food, and so on, when the quantity of food, space and other animals of the same kind are kept at an optimum and other animals of different kinds are excluded from the experiment' (Andrewartha and Birch 1954). It takes into account most of the environmental factors affecting populations, except parasites and predators, and therefore has a meaningful bearing on populations in nature. The innate capacity for increase is directly related to the fecundity, speed of development and mortality of individuals within a population. The physiological aspects of these factors have been discussed previously but their ecological importance must be considered now. The effect of environmental factors on the mortality of animals is often stressed to the exclusion of the effect on fecundity and speed of development, and it is the last two which are often more important in determining the eventual density of the population.

Most invertebrate populations have a large number of environmental factors which affect them in one way or another, but there are usually only one or two of these factors which are important. These can be called 'key factors' (Morris 1959) and it has occasionally been possible to give a quantitative value to key factors for some populations (Varley and Gradwell 1960). Slugs have not received sufficient study to do this and all we can hope to do now is to try and identify the key factors.

Density-independent factors

The most important factor of this type is weather. The effect of weather on the fecundity of slugs is not directly established but it is known that slugs do not lay eggs during very dry or very cold weather. It has not been shown however that individual slugs lay fewer eggs during a year containing a severe winter or dry summer. General metabolism seems to be reduced during unfavourable conditions, but after a long winter slugs may make up for lost time and lay the eggs they would have done if the humidity or temperature had not fallen.

The effect of weather on speed of development is considerable. For example *Milax budapestensis* eggs hatch in about three weeks at 20°C, but take over four months to hatch at 7·5°C. In twelve weeks *Arion hortensis,* when kept at a constant temperature of 20°C, will grow to a weight of 500 mg, but in twelve weeks at 5°C will reach a weight of only 50 mg. Thus, temperature can affect the length of time between one generation and the next. For example, after a short winter most *Milax budapestensis* will hatch in early April; these will grow to about 400–500 mg by the following spring and will then mature and lay eggs —an interval of twelve to eighteen months between generations. But after a long winter many slugs will not hatch until late April or May, and these will not be large enough to mature in the following spring. *Milax budapestensis* is not able to mature during the summer months so that late hatching slugs will take nearly two years before completing their life cycle. The effect of moisture on growth rates of slugs is probably similar to that of temperature. Slugs do not feed when the relative humidity of the microhabitat is less than 100%, and there are probably periods during the summer when surface activity is restricted. In order to calculate the relative humidity of pore spaces in a soil, the water-holding capacity (pF) must be established. The relative humidity of pore spaces falls below 100% at pF 4·2 (for the market garden soil, when moisture was less than 20% of the dry weight of the soil). The soil moisture on the sampling plot fell below this level on only four occasions (moisture was estimated every two weeks) during the twenty-seven months of sampling, but this may have been enough to delay slug growth. There is some evidence (Hunter and Symonds, unpublished) that the very wet summer of 1968 advanced the spring 1969 breeding season. Thus, these slugs were exposed to cold conditions when they were young and vulnerable.

A change in speed of development can have one of two effects on population density. Firstly, it can affect the innate capacity for increase. For example, the fall in capacity for increase of *Milax budapestensis* with a generation interval of twenty-four months instead of eighteen months has an effect equivalent to a 60% mortality at an early stage, or a 60% depression in fecundity. Secondly, a change in generation interval may expose a population to adverse conditions at a vulnerable stage of its life cycle.

The effect of weather on mortality has not been fully established by controlled experiments. It is almost impossible to maintain an artificial temperature or humidity of known value in the immediate vicinity of a slug, because the mucus interferes with established

Ecology

experimental conditions. However, it has been shown that slugs cannot survive for long periods when the relative humidity of the atmosphere is below 90% or when the temperature falls below about $-3°C$. In experiments by Pinder (1969), 50% of *Agriolimax reticulatus* survived a temperature of $-3.5°C$ for thirty-five hours but were killed by $-4.5°C$ in twelve hours; 50% of *Arion hortensis* were killed in twenty-eight hours at $-3.5°C$ and in twelve hours at $-4.5°C$. At $-2.5°C$ slugs lived for three weeks with no mortality. Further experiments, particularly in the field, are required on this subject.

Another important factor, which is not dependent on the density of the population, is the effect of man, both by the use of chemicals for slug control and, indirectly, by altering the nature of the environment during the natural course of cultivating and growing agricultural and horticultural crops. The studies on horizontal distribution of slugs have shown how much the species composition can be affected by man when he changes the environment while cultivating and growing agricultural and horticultural crops. The mechanical action of implements, trampling cows or, on a small scale, gardeners' boots must kill a number of slugs, but the long-term effect on the texture of soil can also be important. Continuous cultivation by farmers or zealous gardeners will compress the soil and reduce the number of spaces for shelter from parasites, predators or extremes of weather.

Finally, a density-independent factor, which is often considered to depend at least partly on the density of the population, is the amount and quality of available food. It is argued that dense populations will reduce the quantity and quality of its own food supply more quickly than a sparse population, but this only applies when the supply of food is less than sufficient for the entire population (i.e., when a shortage leads to competition). With most slug populations, there is a vast surplus of food and the number of animals feeding can have little effect on it. The supply of food for slugs therefore does not depend on the density of the population. On the very rare occasions when there is a shortage of food, it is because of man deliberately keeping the ground fallow. Even this shortage is unlikely to be acute enough to affect the mortality of a population, but it may slow down the growth rate, delay egg laying for a while, or possibly depress fecundity. Experimental evidence on the quantitative food requirements of slugs is scant (see Chapter 3), but speed of development appears to be also dependent on the quality of food. Slugs supplied with poor-quality food, rough grass or straw, for example, will grow more slowly than those fed on potato or succulent green material.

Density-dependent factors

The most important density-dependent factors are predators, parasites and diseases. Very little is known about the effect of these on slug populations, but because slugs are usually nocturnal they may be less liable to depletion by predators. Slugs are, however, regularly eaten by many birds such as starlings or gulls. Collinge (1921) estimated that 6·5% of the total bulk of food of the starlings *Sturnus vulgaris* consisted of slugs and snails, but Dunnet (1956) did not mention slugs as a significant source of food for these birds. Lockie (1956), in his investigation of the food of the jackdaw *Corvus monedula*, the rook *Corvus frugilegus* and the common crow *Corvus corone* does not mention slugs, but Holyoak (1968) found slugs in the gizzards of jackdaws and rooks. Collinge (1924) and Creutz (1963) found slugs to be insignificant sources of food for gulls, but Thorpe and Hope (1908) found thirty slugs in the crop of one gull. Blezard (1967) found occasional slugs in the common and blackheaded gull. Although most of these records do not recognise slugs as being a significant food source for birds it is quite possible that slugs were not noticed in specimens that were examined. Apart from the jaw and, in some species, the shell, slugs are quickly digested and would not be easily recognisable in the gut contents of predatory birds. Flocks of up to 700 common and blackheaded gulls are not uncommon in fields of stubble or following the plough (Vernon, 1970a,b). These could make a significant impact on a slug population. The frequency with which particular slug populations are visited by such flocks has not, however, been established.

Hedgehogs and other small mammals are known to eat slugs, and predaceous insects, such as carabid and staphylinid beetle larvae, have been observed feeding on live slugs. Slugs are often infested with flukes, nematodes, histophagous protozoa or parasitic fungi, but they seem able to cope with a high density of parasites without ill effect. Records of the invertebrates found associated with slugs have been reviewed by Stephenson and Knutson (1966); a summary of this review is given in Table 14. In certain areas (e.g. Hertfordshire) large numbers of *Agriolimax reticulatus* are infested with the larvae of sciomyzid flies, but elsewhere (Northumberland and Cambridge) only a few parasitised individuals were noted in slug populations observed over fairly long periods. The slug mite, *Riccardoella limacum*, is

Ecology

commonly associated with slugs (Turk and Phillips 1946) without causing apparent damage to the host, there is frequently a greater proportion of *Milax budapestensis* infested than *Agriolimax* or *Arion* species (White 1959).

The other major density-dependent factor is competition between animals of the same or different species for resources that both require. The main resources required for the survival of slug populations are food and shelter and as has been described above, there is no evidence that a shortage of these leads to competition.

CONCLUSION

We have seen how the distribution and numbers of a population of slugs varies in space and time, and we have outlined the factors which can be expected to affect the density in nature. Can any of these be identified as 'key factors'?

Firstly there is no evidence that the density-dependent factor, competition for resources, affects the density of slug populations. When a shortage of food or shelter occurs in nature it does not appear to lead to competition between slugs. The effect of the remaining density-dependent factors, parasites and diseases of the egg stage, may be more significant. Also, occasional visits from flocks of birds could significantly affect populations, particularly of the surface-dwelling species. The effect of adverse weather conditions, however, is clear. There can be no doubt that, particularly when shelter is scarce, large numbers of slugs are killed by drought and frost. Weather can also have a less direct but possibly more powerful effect by shortening or lengthening the generation interval, and the effect of this may be more important in determining slug population densities than the direct effect of weather on mortality.

Studies on the spatial distribution of slugs suggest that the intensity of the controlling effect of weather and predators is largely determined by the quantity and type of shelter available. Whether this shelter acts as protection from parasites and predators, or from drought and frost, has not been established.

There seems, therefore, to be three possible key factors which affect slug populations: the action of predators, the direct effect of frost and drought on mortality, and the indirect effect of general weather conditions on generation interval. The relative importance of these factors has not been established, and it will require more population studies before they can be evaluated.

TABLE 14

SPECIES OF INVERTEBRATES ASSOCIATED WITH SLUGS

PROTOZOA		PLATYHELMINTHES	
'Parasite'	'Host'	'Parasite'	'Host'
Concophthirus steenstrupi (Stein)	Agriolimax agrestis Arion ater	Monocercus sp	Arion sp
Paraglaucoma limacis (Warren)	Agriolimax agrestis	Dicrocoelium dendriticum (Rudolphi)	Arion subfuscus Limax tenellus Agriolimax reticulatus
Colpoda aspera Kahl	Agriolimax agrestis	Brachylaema nicolli (Witenberg)	Agriolimax agrestis
Colpoda steini Maupas	Lehmannia marginata Agriolimax reticulatus	Panopistus pricei Sinitsin	Agriolimax agrestis Agriolimax laevis
Tetrahymena pyriformisi (Ehrenberg)	Agriolimax reticulatus	Brachylaema virginiana (Dickerson)	Ariolimax columbianus
Tetrahymena rostrata (Kahl)	Agriolimax reticulatus	Brachylaema sp	Milax sowerbyi Agriolimax reticulatus Arion lusitanicus Arion ater ater
Tetrahymena limacis (Warren)	Agriolimax reticulatus		
Pfeifferinella impudica Léger et Hollande	Limax marginatus		
Isospora incerta Schneider	Limax cinereoniger	Davianea proglottina (Daviane)	Agriolimax agrestis Agriolimax reticulatus Agriolimax caruanae Milax budapestensis Milax sowerbyi Limax cinereoniger Limax flavus Arion ater ater Arion hortensis Arion fasciatus Arion intermedius
NEMATHELMINTHES			
Leptodera appendiculata (Schneider)	Agriolimax agrestis Arion ater		
Leptodera flexilix (Dujardin)	Limax sp		
Leptodera angiostoma (Dujardin)	Arion ater		
Strongylus sp	Limax maximus Limax marginatus	Anomotaenia subterranea Cholodkowsky	Arion ater rufus
Oxyurus sp	Limax maximus Limax marginatus		

Parasite	Hosts	Parasite	Hosts
(Railliet)	*Arioumax columbianus* / *Limax flavus* / *Arion fasciatus*	*Geoplana septemlineata* (Hyman)	*Agriolimax reticulatus*
		INSECTA	
Cosmocercoides sp	*Agriolimax agrestis*	*Sarcophaga melanura* (Meigen)	*Arion hortensis*
Cosmocercoides dukae (Holl)	*Agriolimax* sp	*Carabus violaceus* (L)	*Arion hortensis* / *Milax gagates* / *Agriolimax reticulatus*
Alloionema appendiculata Schneider	*Arion ater ater*	*Scaphinotus interruptus* (Menetries)	No species mentioned
Mermis nigrescens Dujardin	*Agriolimax agrestis*	*Lampyris noctiluca* (L)	*Arion ater ater* / *Arion subfuscus* / *Arion hortensis* / *Arion fasciatus*
Muellerius capillaris (Müller)	*Milax sowerbyi* / *Agriolimax reticulatus* / *Agriolimax laevis* / *Agriolimax agrestis*	*Phausis spendidula* (Le Conte)	*Arion ater ater* / *Arion subfuscus* / *Arion hortensis* / *Arion fasciatus*
Limaconema sp	*Arion* sp / *Limax* sp	*Tetanocera plebeia* Loew	*Agriolimax laevis*
Hexamermis albicans (Seibold)	*Agriolimax agrestis*	*Tetanocera valida* Loew	*Agriolimax laevis*
Angiostrongylus cantonensis (Chen)	*Limax* sp / *Agriolimax laevis* / *Veronicella leydigi*	*Tetanocera clara* Loew	*Pallifera dorsalis* / *Philomycus carolinianus*
Cystocaulus ocreatus (Railliet et Henry)	*Agriolimax reticulatus*	*Tetanocera elatata* (Meigen)	*Agriolimax reticulatus* / *Arion fasciatus* / *Arion intermedius* / *Milax budapestensis* / *Milax sowerbyi* / *Limax tenellus* / *Limax flavus*
Anaflaroides rostratus (Gerichter)	*Agriolimax laevis*		
Rhabditis cf. *lambdiensis* (Maupas)	*Agriolimax reticulatus*		
Parogrolaimus sp	*Agriolimax reticulatus*	*Abax striola* (F)	*Agriolimax reticulatus*
Diplogaster sp	*Agriolimax reticulatus*	*Feronia melanaria* (Illiger)	*Agriolimax reticulatus*

8

SLUGS AS PESTS

Introduction

The word 'pest' comes from the Latin in which it meant what we now call bubonic plague. In English, the word has been transferred from the disease to the organism that causes it, transmits it, or even in general 'something troublesome or noxious' *(Oxford English Dictionary)*. Slugs can be regarded as pests when they are in a wheat field at the time of drilling, but they are not pests in the same field six months later, or on an adjacent roadside at any time of year.

Slugs damage a wide variety of crops. As with many other agricultural pests, there is very little information on the extent or the value of damage they do, and it is very difficult to estimate where they stand in the order of importance as farm and garden pests. Very little attention has been given to the pest status of slugs mainly because they are so difficult to control and because their damage tends to be localised and unpredictable.

Slugs are of economic importance in many parts of the world. Many Ministries of Agriculture publish official leaflets or include chapters in official bulletins on their control (Anon., 1951; 1953; 1959; 1965; 1966; 1967; 1969; Wilkinson 1964; Fenner 1962; Anderson & Nilsson 1967; Braithwaite 1961; Jenkins 1960). An attempt to ranks slugs as of major or minor importance (Hunter 1969a) was not helpful because of the intrinsic difficulties of comparing damage in a country such as Sweden with that in a country like the USA where pests in general are much more important. From replies from officials in Ministries of Agriculture, considered together with official leaflets and bulletins, it was concluded that slug damage on a world scale was of considerable significance. Although there were several pest groups in each country which could be considered of

Slugs as pests

greater importance than slugs, few groups have such a wide geographical distribution or affect such a wide range of crops.

Slugs are known to be important pests of potatoes in the British Isles. For example, it has been estimated by Strickland (1965) from Potato Marketing Board records in England and Wales that they are the third most important pest of this crop. Aphids, mainly through their role as vectors of viruses, are estimated to result in a crop loss of about 34,400 acre equivalents (out of a total 416,000 acres) of maincrop and second early potatoes (i.e., 34,400 extra acres are required to make up for the loss in yield due to aphid damage). Potato cyst eelworm is estimated to result in a loss of 10,150 acre equivalents and slugs a loss of 3,300 acre equivalents. Slugs can therefore be said to cause a loss of about 36,000 tons of potatoes each year—the average annual consumption of about 400,000 people.

Slug damage to potato tubers is distinctive and easily diagnosed (Fig. 53). A hole in the surface of the tuber leads to a chamber hollowed out beneath. The type of hole can be distinguished from wireworm damage where a small hole is bored straight into the tuber, and from cutworm damage where the hole is wide on the surface and gradually narrows as it penetrates deeper.

Slugs are one of the three most important pests of winter wheat. Strickland (1965) has estimated that in England and Wales they are responsible for the loss of 41,000 acre equivalents of wheat (out of a total of 2,172,000 acres) compared to losses of 36,000 from wireworm, 30,000 from wheat bulb fly, 14,000 from rabbits and hares and 13,000 from cereal root eelworm. It is a little misleading to assess the damage of most of these pests in terms of 'acre equivalents', since an acre of winter wheat destroyed is not totally lost: there is usually time to redrill the land with spring wheat. The lower yield and the cost of new seed and labour for re-drilling results in a loss to the farmer of only about one-quarter of the loss he would suffer if the crop was completely wasted. The acreage of wheat actually re-drilled because of slug damage may be as low as 25,000 acres in an average year (Hunter 1969b).

There are two types of slug damage to wheat. Firstly, and most important, is grain hollowing (Fig. 57), where the germ of the grain is eaten out shortly after drilling, thus preventing germination. Again, the hole hollowed by the slugs tends to be wider inside the grain than on the surface. Secondly, slugs can graze or shred the wheat plant after it has germinated. The shoots and leaves appear above the soil surface, but they are eaten away as they grow (Fig. 56). Usually only

the succulent parts of the leaves between the veins are eaten, so that
the plant tends to have a ragged appearance. This distinguishes the
damage from that of birds or small mammals where the leaf is bitten
cleanly off. The wheat plants are usually able to grow quickly enough
to withstand leaf damage and there is some evidence to show that a
certain amount of leaf damage stimulates the plants to tiller, thereby
increasing yield. Occasionally, however, the weather is just warm
enough at night (above 2°C) for slug activity, but is not warm enough
(about 5°C) during the day for the plants to grow. On these occasions
slugs can graze the plants enough to retard their growth or even kill
them completely. Spring wheat is not so susceptible to slug damage
because it germinates and grows more quickly, giving the slug less
time to damage it.

Damage by slugs to wheat is becoming increasingly important as
farming progresses towards fewer cultivations and lower seed rates.
There have also been recorded instances (e.g., Johnson 1968) of crops
which have been severely damaged when they have been drilled by slit
seeding after the field has been treated with paraquat.

Slugs also damage newly established grass leys by grazing the
newly emerged leaves and a number of crops are reported to have
failed because of slugs each year in the west of England and Wales.
Crops of barley and oats are also occasionally reported to have failed
due to slugs grazing the shoots just below the soil surface: this can
easily be confused with similar damage by wireworms.

Slug damage to sugar beet is also becoming more important. The
traditional method of growing sugar beet was to drill a large number
of seeds and remove the unwanted seedlings later. This is now being
replaced by monogerm seed drilled at low rates so that there are
fewer seedlings. Previously, the loss of some seedlings was unimportant but if precision drilled seedlings are destroyed there are no
immediately adjacent plants to take their place. There are no estimates of the acreage of sugar beet damaged by slugs, but it is unlikely
to be large compared to that of wheat and potatoes.

There are no estimates of the extent of damage to horticultural
crops, but it is possible that this exceeds the agricultural damage.
Slugs have been recorded as causing widespread loss of brusselssprouts, cabbage, lettuce, bulbs and flowers. These crops are
intensively grown and if in good condition very profitable. Often the
yield of the crop is not affected, but the quality is reduced so that the
product is unsaleable or reduced in value. Thus, damage can result in
considerable losses to the grower. This is particularly so with brussels-

Fig. 53 (ABOVE) Slug damage to potatoes.
  a. External view of a potato holed by a slug.
  b. The same potato cut open to show the cavity which is enlarged below the surface.

Fig. 54 Slug damage to potato plants. Leaf damage (Maris Piper).

Fig. 55 (ABOVE) Two potato plants growing adjacent to each other. The Maris Piper (on the left) has been decimated. The Majestic plant (on the right) has received little damage.

Fig. 56 (BELOW, LEFT) Wheat grains hollowed by slugs.

Fig. 57 (BELOW, RIGHT) Wheat leaves shredded by slugs grazing between the veins.

Slugs as pests

sprouts, where a high proportion of the acreage grown is affected. The sprout buttons are holed by slugs and have to be rejected. Slug damage may cause the 'blowing' of sprouts, where the leaves of the button fail to close into the required tight shape. A considerable effort is also required before marketing other horticultural crops to ensure that the produce is not contaminated by the presence of slugs. The frozen pea crop is particularly susceptible to this type of contamination. Peas are often vined at night when slugs are active high in the foliage. The peas (and slugs) are transported to the freezing plant within one or two hours of vining and frozen slugs are not easily separated from frozen peas in the sorting machinery (Ensor and Wharton, personal communication).

The economic importance of slugs as intermediate hosts of parasites of domestic and wild animals has received less attention than their role as crop pests. Parasitologists do not appear to have realised that slugs are often more numerous and more widely distributed than the accepted snail intermediate hosts of parasites. For example, there appears to have been no comprehensive investigation into the possibility of slugs transmitting *Fasciola* species (the liver flukes), the most important parasites of domestic animals, which are usually carried by *Lymnaea* snails. However, the sheep lungworm *Muellerius capillaris* is known to have *Agriolimax reticulatus, A. agrestis, Arion ater, Limax maximus, L. flavus* and *Milax sowerbyi* as intermediate hosts (Hobmaier and Hobmaier 1929, Williams 1942, Beresford-Jones 1966). Williams (1942) maintained that *Agriolimax reticulatus* was the most important intermediate host in the transmission of *Muellerius capillaris* larvae. Other parasites of economic importance have slugs as their intermediate host (Lapage 1956, Cheng 1964). *Davainea proglottina* (Cestoda) and *Syngamus trachea* (Nematoda) are important parasites of domestic fowl (Brown, 1933). *Dicrocoelium dendriticum* (Trematoda) parasitises domestic herbivores. *Brachylecithum orfi* (Trematoda) is a parasite of grouse (Kingston and Freeman 1959).

Slugs are also intermediate hosts of the parasites of wild life shown in the table on p. 142.

The effect of the parasites on the slug has received little attention but most intermediate hosts can withstand a high degree of parasitisation without ill effect. If the slug is not killed before the parasite has reached the infective stage, it is not an effective intermediate host. Foster (1958a,b) found that the Brachylaemid parasite of small mammals caused necrosis in the kidney of the slug intermediate hosts.

Host	Parasite	Reference
Amphibians	*Brachycoelum obesum* (Trematoda)	Cheng (1964)
Rodents	*Postharmostomum helicis* (Trematoda)	Malek (1962)
Birds	*Leucochloridium macrostoma* (Trematoda)	Cheng (1964)
Birds	*Brachylaima nicolli* (Trematoda)	Malek (1962)
Small mammals	*Brachylaima virginianum* (Trematoda)	Malek (1962)
Short-tailed shrew	*Panopistus pricei* (Trematoda)	Malek (1962)
Short-tailed shrew	*Entosiphonus rhomboideus* (Trematoda)	Malek (1962)
Shrews	*Anomotaenia subterranea* (Cestoda)	Prokopič and Ždarska (1958)
Cat	*Aelurostrongylus abstrusus* (Nematoda)	Blaisdell (1952)
Rat	*Angiostrongylus cantonensis* (Nematoda)	Cheng (1964)
Bobcat	*Anafilaroides rostratus* (Nematoda)	Malek (1962)

Milax sowerbyi could withstand high infestation without increased mortality but the food intake and egg production of infested *Agriolimax reticulatus* was reduced and, with a severe infestation, they died earlier.

Slugs can also transmit plant diseases. Cabbage leaf spot (Hasan and Vago 1966) and downy mildew of lima beans (Wester, Goth and Webb 1964) can be carried from plant to plant after the spores have passed unharmed through the slug gut. Slugs may also be implicated in the transmission of potato blight (Webley, personal communication). The predilection of slugs for fungal foods may increase the attractiveness of diseased plants and thus increase the probability of disease being spread by slugs.

METHODS OF SLUG CONTROL

Until recently control of a pest was often taken to require its eradication. Nowadays we take a more realistic view, and we mean by pest control the regulation of the environment of the crop so that the

Slugs as pests

damage does not reach an economic level. We are quite content to share our cabbages with caterpillars, provided that the cost of their damage or nuisance is not greater than the cost of controlling the pest. It is often the nuisance value of the pest which is the important factor. Slugs do not eat a large weight of potatoes but they make a very much larger weight of potatoes unsightly and unsaleable. Nowadays the risk that half a slug may be found in a packet of frozen food is enough to have serious economic effects.

Pest control is usually, but not always, achieved by reducing the numbers of the pest. But higher populations of pest species are quite acceptable if we can modify the planting date of the crop, change the variety or in some other way make the crop unavailable to the pest. Most of the known methods of pest management are of some value in the control of slugs and we will describe them under the headings of biological, cultural and chemical control.

Biological control

Biological control has been taken to mean the control of pests by the introduction and management of their natural enemies. This method has been used with some success for the control of pests in clearly defined and isolated areas (e.g., in glasshouses), or in environments where the pest has been newly introduced. Most natural enemies of slugs probably do not have a significant effect on slug populations, however, and those that may have an effect (birds) can not be 'managed' for pest control purposes. The introduction of flocks of ducks has been recommended for slug control (Curtis 1860), but this seems an impracticable method today.

Cultural control

Under this heading we can consider all of those modifications which can be made to the husbandry of the crop to reduce slug damage.

The most effective of these modifications is to the cultivation that the land receives before and during the growing season. The effect of various cultivations on populations of *Agriolimax reticulatus, Arion hortensis* and *Milax budapestensis* was assessed during the population study in the market garden already described in Chapter 7 (Hunter 1967). When the ground was broken up into a fine tilth by rotovating three times, the populations of *Agriolimax reticulatus* and *Arion hortensis* were reduced to about a quarter and one-third of their previous densities. Numbers of *Milax budapestensis* were not reduced significantly, presumably because this species lives deeper in the soil and has a tough skin which is not easily damaged by mechanical action. Populations were estimated immediately before and after the cultivations so that reductions in numbers could be attributed directly to the mechanical action of the implement. Another observation showed that thorough cultivation, followed by compaction of the soil and establishment of a grass sward which was then frequently mown for three summer months, almost eradicated populations of *Agriolimax reticulatus, Arion hortensis* and *Milax budapestensis*. This more pronounced effect was presumably due to the absence of large cracks and spaces in the soil so that slugs could not find shelter from drought or predators. The absence of soil spaces also inhibits slug movement and this can reduce the amount of slug damage to agricultural crops, particularly winter wheat. If there are no spaces in the soil, the slug is unable to move through it to reach the growing seed. The effect of this can be seen in many wheat fields where there is a good stand of the crop around the edge of the field but the centre is severely thinned by slugs. The headland is compacted by implements passing over it more frequently than in the centre and the slugs are unable to damage the grain drilled in this part of the field. Occasionally, even the wheel marks made by a tractor passing over the field after the crop has been drilled can be distinguished from other parts of the crop. The soil has been compressed by the tractor wheels and the slugs are unable to reach the seed growing in this narrow area.

There are, however, two major disadvantages in giving extra cultivations to fields at risk from slug damage. Firstly, cultivations are expensive: it is frequently cheaper to treat with chemicals for slug control than to give an extra pass with a cultivating implement. Secondly, on the heavy land where slugs are commonest, too many cultivations can lead to over-compaction of the soil and reduce plant root growth. In any case, it is frequently impossible for machinery to travel on heavy land during wet weather.

Another cultural method of controlling damage by slugs is to grow varieties of the crop that are less susceptible to attack. As we have shown in the section on feeding behaviour, slugs feed on a wide range of food material, but when given a choice are remarkably discerning in their feeding habits. Slugs have not been shown to distinguish between varieties of most crops, but they are known to have distinct preferences for certain varieties of potatoes (Gould 1965, Winfield, Wardlow and Smith 1967, Hunter, Symonds and Newell 1968). When different varieties are grown alongside each

Slugs as pests

other the slugs feed much more actively on some varieties than on others. It has been possible to construct a table of the major differences in varietal susceptibility:

Very High Susceptibility	High Susceptibility	Medium Susceptibility	Low Susceptibility
Maris Piper	King Edward	Majestic	Pentland Falcon
Ulster Glade	Record	Pentland Dell	
	Pentland Crown		

This list, of course, only refers to maincrop varieties since the early varieties are usually lifted before damage begins. It is not known why the varieties differ in their susceptibility to slug attack but differences have been noted in both tubers and leaves (Fig. 55). Biochemical rather than physical factors such as skin thickness or size and depth of tubers are therefore most likely to cause this difference, though the latter factors may have a secondary influence. It is also unclear whether the differences in susceptibility are due to variable attractiveness of root or tuber diffusates from a distance, or to differences in taste once feeding has begun. It is quite possible that the slugs are not attracted to the very susceptible varieties, but are stimulated to feed more when they have come into contact with them. It is clear, however, that farmers would often be wise to accept a reduction in potential gross margin by, say, growing Majestic instead of King Edwards.

A further cultural method of controlling damage to potatoes is lifting the crop early in the season. It has been noted in the east of England that damage can increase tenfold between the beginning of September and the beginning of October (Hunter and Symonds, unpublished). This is probably due to the interaction of four factors. Firstly, the amount of food on the surface declines in the autumn: the potato leaves die off as the tubers mature and weeds become old and unpalatable. Secondly, the numbers of *Agriolimax reticulatus* and *Arion hortensis* and the size of other species are increasing during the autumn. Thirdly, the weather is usually moist and more favourable to slug activity. Fourthly, biochemical changes within the tuber as it matures may render it more susceptible to slug damage. The relative importance of these factors in the increase of susceptibility has not been established.

Slug damage can be reduced by planting the crop when the susceptible period will not coincide with weather favourable to slug activity. This may have particular relevance to wheat, although there are a number of conflicting factors involved. If wheat is drilled early in the autumn, the slugs are still active and will do more damage than if drilling is delayed until the onset of cold weather. On the other hand if the weather does not become cold in late November and December, slug populations are higher (after the second generation of *Agriolimax reticulatus* is well established). In addition, later in the autumn the temperature may be low enough to delay germination, but not low enough to depress slug activity. Yet a further consideration is that later drilling will render wheat more susceptible to attack from wheat bulb fly. If slug attack is certain (it rarely is), the farmer may be wise to delay drilling until the spring when slug populations are lower and when the crop can germinate quickly.

We have already seen from the ecology of slugs that shelter is important in the establishment and maintenance of dense slug populations. It is therefore important that any residue from the previous crop is removed from the surface as soon as possible. Straw or trash from vining peas can give ideal cover for slugs in the late summer but its removal will allow the soil surface to become too dry for the slugs to survive. It has also been noted that ploughing in chopped straw to improve the structure of the soil leads to an increase in slug damage. It is unfortunate that reduction in slug damage on heavy soil is often incompatible with improving the structure of these soils.

Chemical control

Chemical control depends on the application to the crop of a chemical which kills or incapacitates the pest or protects the crop from damage. We usually include under this heading the application of repellents to the crop, but not the use of fertilisers which can counteract damage by producing additional growth.

The first chemical to be used extensively for slug control was metaldehyde. Gimingham (1940; Gimingham and Newton 1937) reported that this chemical had been accidentally discovered to have slug-killing properties and it was developed on a wide scale for commercial use. None of the organo-chlorine or organo-phosphorus compounds, that were developed on a wide scale for use against insects, were reported to be active against slugs. A third group of insecticides, the carbamates, was then developed. Ruppel (1959)

Slugs as pests

reported that the first carbamate to be widely used, carbaryl, was toxic to slugs. Since then a number of carbamates have been shown to have slug-killing properties.

The major difficulty in assessing the relative toxicity of chemicals against slugs is that screening methods are laborious and, compared to insecticide screening, inaccurate. Most methods for examining the relative toxicity of stomach poisons depend on the slugs voluntarily ingesting test substances. Stringer (1946), Webley (1965), Ruppel (1959), and Crowell (1967) tested various chemicals or various concentrations of metaldehyde by presenting baits to confined slugs and noting the effect of these after the slugs had eaten from them. This method, however, does not separate the effects of 'attractiveness' and 'toxicity'. Also, no indication is given of the weight of the chemical required to kill the slug.

Cragg and Vincent (1952) attempted to measure the toxicity of metaldehyde by forcing known quantities down the alimentary canal of *Agriolimax reticulatus*. Finely ground metaldehyde was mixed with water and then introduced into the buccal cavity using a microsyringe with a fine glass delivery tube. It was found that only 0·2 ml of suspension could be injected without the slug regurgitating, and the results from the experiments were too erratic to permit accurate assessment of the lethal dose. Henderson (1969a) developed this technique by mixing the chemical with agar jelly and injecting this into slugs anaesthetised with carbon dioxide. By this method up to 10 ml of mixture can be injected into the slug. The slugs are then passed through a foam rubber mangle to squeeze the chemical past the circumoesophageal nerve ring. Henderson calculated the 'median lethal dose' (the amount required to kill 50% of the treated slugs) was 130 μg of copper sulphate, 85 μg of metaldehyde, or 23 μg of sodium pentachlorophenate per slug. The method has been used by Stephenson (in Johnson 1967) to test the activity of ioxynil and by Hunter and Johnston (1970) to assess the activity of a number of carbamates. The various workers found that results were extremely variable depending on the conditions in which slugs were kept immediately preceding and following treatment. Nevertheless, it has been established that the oral toxicity of metaldehyde is over twice that of copper sulphate, and the carbamates have a very wide range of toxicities. Some carbamates are over five times as toxic as metaldehyde and some are much less toxic.

Laboratory methods for assessing the toxicity to slugs of contact poisons have been described by Henderson (1969b). It is not possible to apply known doses of test substances directly on to the slug because irritant materials are sloughed off in the exuded mucus. Allowing the slug to crawl over treated surfaces gives erratic results because the distance moved by the slug affects the amount of chemical picked up; a substance which quickly immobilises the slug without killing it will not be administered in as high a dose as a substance which allows the animal to travel further.

A total immersion method has none of these disadvantages. Slugs can withstand total immersion in water at 10°C for over five hours. In the screening tests, slugs were immersed in a solution of the chemical with water for periods of one hour. For substances such as metaldehyde, which do not dissolve easily in water, an additional test was made. The chemical was thoroughly mixed with a finely ground talc and the slugs immersed in this mixture; they were washed with a jet of water immediately after removal. As with the stomach poison screening method, extreme care was taken before and after treatment to ensure that conditions were absolutely standard. The data from these immersion tests indicated that the median lethal dose for ioxynil was 8·3 ppm, copper sulphate approximately 70 ppm, sodium pentachlorophenate 22 ppm, and metaldehyde 4,822 ppm. Using the dry contact method, metaldehyde (42,370 ppm) was much less toxic than copper sulphate (2,027 ppm), but the method may have introduced a complicating fumigant effect.

Chemicals for pest control can be applied in a number of formulations. Firstly, they can be dissolved or suspended in water and applied as a spray. This gives an even but not very persistent distribution of chemical over the surfaces. Although some sprays are taken up by plants ('systemic') and others adhere to the foliage for long enough to be eaten by grazing pests, the majority function primarily as contact poisons. Slugs are very difficult to kill by contact poisons because they are able to slough off any substance which is noxious. Metaldehyde can be applied as a spray (usually 4% of active material) but gives unsatisfactory results unless used at a very heavy rate when the slugs are active. These sprays will not persist on the soil or on foliage when the weather is wet and it is only at this time that slugs are active. Metaldehyde sprays have been found to provide some control of slugs on crops in easily controlled habitats (e.g., glasshouse crops or forced bulbs) but they are not recommended on an agricultural scale. Some of the more toxic carbamates are probably active enough to be applied as sprays.

Secondly, poisons may be mixed with a carrier which is attractive

Slugs as pests

to slugs (for example, bran or wheat meal) and offered to the slug in the form of a bait. Copper aceto-arsenite ('Paris green'), can be finely ground and mixed with bran to form a slug-killing bait. When this bait is broadcast at the rate of 28 lb per acre it will provide some control of slugs. This mixture, however, is dangerous to handle and does not give good control. The use of Paris Green may be justified when leather jackets (the larvae of *Tipula* species), which it also kills, are pests on the same field.

Metaldehyde can be formulated and used in this way but there are several proprietory brands of ready mixed baits available. These baits are compressed into pellets for distribution over the ground. The pellets vary in metaldehyde concentration, size and persistence, and it has been shown that all three of these factors have a bearing on effectiveness. Trials (Hunter and Symonds, unpublished) have shown that pellets containing 6% metaldehyde give significantly better control than those containing 3, 4 or 5% metaldehyde. If the metaldehyde content falls below about 5% most of the slugs eating from pellets are able to recover, especially when the weather on the following day is warm and moist.

The size of pellets has a bearing on the distribution rate (Hunter and Symonds, 1970). A given weight of small pellets will clearly have more killing points (pellets) per unit area than the same weight of large pellets. The probability that a slug will encounter a bait pellet can be estimated theoretically by assuming that slugs move at random over the ground where pellets have been randomly scattered. Where P = the probability that a slug will encounter a pellet, x = the distance over which each pellet 'attracts' a slug, y = the distance moved by a slug in a unit of time and A = the area in which each pellet is placed (i.e., if there were one pellet in each ten units area, A = 10) then:

$$P = 1 - e^{\frac{-2xy}{A}} \quad (e = \text{base of Natural logarithms})$$

Thus, the efficiency of the bait will depend on the 'attractiveness' of the pellets and the density of their distribution. In cold or dry weather the efficiency will decline because of lower slug activity. Although the various parameters for this model have been incompletely defined, a series of observations on the movement and mortality of *Agriolimax reticulatus* on plots with various densities of metaldehyde and carbamate pellets demonstrated that the model could give some indication of optimum pellet density. When the observations were made, the slugs were covering about 85 cm while they were responsive to stimuli indicating the presence of food, and the pellets were estimated to attract slugs over approximately 4 cm: using these parameters, the optimum distribution of pellets was found to lie between 10 and 20 cm apart. This estimate, however, was for the most active slug, *Agriolimax reticulatus,* and the experiments were conducted on a closely mown lawn where the slugs could move easily. Other species of slug covering more difficult terrain would not move so far. Thus, the most effective control would be obtained by pellets about 10 cm apart.

It is important that any reduction in the size of bait pellets should not lead to too much loss in persistence. The frequency of nocturnal excursions by slugs on the soil surface has not been studied fully but it will depend on the weather and on the availability of underground food. It can be assumed, however, that many slugs, particularly those that spend the day deep in the soil, do not come to the surface every night. Stephenson (in Johnson 1964) has shown that in mild moist weather most *Agriolimax reticulatus* will visit the surface during any seven-day period.

Control of many pests and diseases of wheat, e.g., wheat bulb fly (Gough and Woods 1957), has been achieved by coating ('dressing') the grain with a chemical which is taken up by the plant, protecting it from later attack. Seed dressings have been tested for slug control (Gould 1962) but so far with little success. It was not possible to load a large enough coating of metaldehyde on to wheat grains in order to reduce significantly their susceptibility to attack: even loadings of ten ounces of metaldehyde per bushel of seed (the usual loading of seed dressings is two to three ounces per bushel) gave little control. Some control was obtained with five to ten ounces per bushel of copper oxychloride but the effect was limited and considerable loss of seed still occurred. These very heavy loadings of seed dressings alter the texture of the grain surface and give rise to irregularities in the flow of seed through the drill.

Again, however, there is evidence that some of the carbamates may be active enough to give control as seed dressings. The ideal solution would be to combine a seed dressing for slugs together with a dressing for the control of wheat bulb fly. Unfortunately, no carbamate so far produced is persistent enough to protect wheat from wheat bulb fly which attacks the plant three to five months after drilling.

INTEGRATION OF CONTROL MEASURES

It has already been mentioned that pest control does not necessarily imply complete eradication of the pest. It does mean, however, that changes are required in the habitat of the animal in the crop (sometimes called 'agro-eco-system'). It is very important that these changes should be related to the general biology, distribution and behaviour of the pest and with methods of crop husbandry. 'Integrated' control has been defined as 'a pest management system that, in the context of the associated environment and population dynamics of the pest species, utilises all suitable techniques and methods in as compatible a manner as possible and maintains the pest populations at levels below those causing economic injury' (Smith 1967). This means that control measures should be used only where they are absolutely necessary, only when they have the optimum effect on the pest, and only when they are applied in such a fashion that they do not have an undesirably detrimental effect on beneficial organisms and environmental factors in the habitat. The words 'undesirably detrimental' and 'beneficial' should be, but are frequently not, defined in their widest possible sense. A reduction in earthworm numbers for example may not have a significant economic effect but in the long term may lead to a deterioration of soil structure. Similarly, song birds can and should be regarded as beneficial.

Where to apply control measures

It is possible to predict the place and time of attack of many insect and invertebrate pests over fairly wide areas. Warnings of impending attacks by pea moth, pea midge, carrot fly, and other insect pests are given each year in England and Wales. Once the practical difficulties of making observations over major areas are overcome, it is still easier to predict the probability of attack by migrating insects, such as the desert locust *Schistocera gregaria,* because their flight path is often known and warnings can be given some time in advance. Fairly accurate predictions can be made of the level of attack by eelworms (for example, *Heterodera rostochiensis,* the potato cyst eelworm, and *Heterodera schachtii,* the beet eelworm) by estimating the density of populations before the crop is planted.

Slug damage, however, is not easy to predict because the slugs themselves are localised and the extent of damage depends on the weather and cultural factors. Extensive damage can be expected on fields where slugs have previously been noted, particularly after crops which have provided cover during the summer months (Gould 1961). Research on the effect of weather on slug populations (Chapter 7) indicates that heavy damage can be expected in years following a wet summer or a mild winter. However, more information is required of the effect on slug populations of preceding weather before we can make predictions of this type with any confidence.

Attempts have been made to predict damage on a field scale. Duthoit (1961, 1964) devised a 'population estimation method' in which wheat grains were placed on the soil surface and the numbers of these damaged by slugs within one week were taken as a measure of the population density. Various workers have developed this as a method for predicting damage to winter wheat. A major drawback of Duthoit's method was that mice and birds frequently interfered with the baits before a record of slug damage could be made. This can be overcome by enclosing the grains in polythene tubing so that the slugs can get to them but birds and small animals are excluded. Disposable petri dishes can also be used if they have holes bored into them large enough for slugs to gain access. Preliminary trials (unpublished) have shown that, if they are placed in the field some time before drilling, a heavy attack by slugs on these grains will indicate an attack to the following crop. For potatoes, a similar method of putting small samples of the main crop down in the field before drilling can be used to predict impending attack. Maris Piper or King Edward tubers are 'planted' about three inches under the soil surface about a month before the crop is planted. These are left for about a week and if most of them are damaged by slugs within that time, damage to the main crop can be expected. Neither of these prediction methods has been fully calibrated, and records over a number of years under different conditions of weather, soil type, etc. will be required before a conclusive evaluation of them can be made.

Timing of control measures

In any battle, a useful principle is to hit the enemy hardest where he is weakest. Similarly, in pest control we try to reduce populations at the weakest point of their life system. The natural growth of a population follows a sigmoid curve with a low initial rate of growth gradually increasing in a logarithmic fashion until density-dependent controlling factors begin to slow down the rate of increase. The aim is to keep the population density down to the lower part of the curve so that subsequent increase in numbers is slow.

This is not always possible in slug control because low densities of

Slugs as pests

slugs are not easy to detect. In any case, farmers and growers cannot always justify 'insurance' measures for slugs when they are not certain that populations will build up at some time in the future. Thus, treatment is usually applied only when slug populations have already reached fairly high densities.

If it is not possible to reduce populations when they are sparse, farmers may be able to apply control measures before the breeding seasons. In Northern Europe, the breeding season of the main pest species do not coincide: *Agriolimax reticulatus* breeds in both spring and autumn, *Arion hortensis* breeds in mid-summer, and *Milax budapestensis* throughout late winter and spring. Nevertheless control measures applied in early spring will reduce the population before the spring breeding season of *Agriolimax reticulatus* and the summer breeding of *Arion hortensis*; populations of *Milax budapestensis* will also be reduced before the spring eggs are laid. Population densities naturally tend to be low after the winter, so that it is easier to reduce numbers to levels where subsequent rate of increase is low.

A second method of integrating the timing of control measures into the agro-eco-system is to ensure that these measures are compatible with crop husbandry methods. For potatoes, the spring control measures coincide well with crop management. If a bait is applied during the growing season, many slugs will remain underground, feeding on potato tubers which they naturally prefer to poisoned baits. It is also impossible to incorporate a slug-killing chemical into the soil during the growing season. A spring treatment, before the seed potatoes are planted, allows both the incorporation of chemicals and the application of baits without competition from other food sources. Particularly with baits, however, the treatment should not be applied before the temperature has risen enough for the slugs to become active.

Spring application of control measures is compatible with the management of other crops such as sugar beet, peas and brassicas which are planted in the spring. It is not so compatible with the husbandry of winter wheat which is drilled in the autumn. Cereals are often grown for several successive years and there is frequently a crop already growing on the land in the spring. In first year crops, slug densities may only build up to detectable levels during the preceding summer. In this case, treatment must be applied in the autumn but should be as early as possible. The main species to attack winter wheat is *Agriolimax reticulatus* and it is clearly an advantage to reduce the population before the autumn breeding season of this species. Since most damage occurs shortly after drilling, it is also

F

important to ensure that population densities are reduced before this time. It is not yet established whether it is better to treat before or after the stubble of the previous crop is ploughed, but it is known that slugs are more active on the surface after ploughing (Duthoit 1964). Thus, surface baiting, for example, should be more efficient after ploughing. The most efficient control for slug damage to winter wheat would be a seed dressing which, of course, would take effect at drilling time, but other control measures should be applied as far as possible before drilling.

Hazards from misuse of control measures

Slug control is not without hazard to other animals, particularly since the most widely used method of chemical treatment is poisoned bait. Baits are readily eaten by a wide range of soil-dwelling animals and wild life. A number of unconfirmed reports (United Kingdom Ministry of Agriculture) have implicated metaldehyde baits in the deaths of domestic animals and game birds. Metaldehyde is not as poisonous to mammals as many other pesticides. This chemical has a median lethal dose for dogs of about 500 to 600 mg of chemical to one kilogram of body weight; thus, it is clear that the affected animals had eaten large quantities of 3% metaldehyde bait. The carbamates are much more toxic and have a range of toxicity from 100 mg/kg to one mg/kg. It is clear that the most toxic carbamates present considerable hazards to wild life, and the users who handle them. It should be possible to use the less toxic carbamates (a range of mammalian toxicity of 30 to 100 mg/kg) as baits if they are formulated so that they are unattractive to wild life. There is some evidence (Bayer Chemicals Ltd., unpublished) that baits made into small blue pellets are not picked up by birds. These pellets are, however, readily eaten by field mice, voles, and other small rodents.

A further hazard of carbamate use for slug control is created by their toxicity to soil invertebrates. Many of these, such as carabid and staphylinid beetles, are predators of slugs. It has already been shown (Chapter 7) that these predators are unlikely to have a significant effect on slug populations but they are also predators of many other pests (e.g., cabbage root fly). Carbamates also kill earthworms, and indiscriminate depletion of earthworm populations may in the long term prove detrimental to soil structure. There are conflicting reports of the amount of earthworm depletion resulting from treatments with carbamate baits. Field trials using these baits have resulted in estimated reductions in earthworm populations of between 4% and 30%.

REFERENCES

Abeloos, M., 1943. Effets de la castration chez un mollusque *Limax maximus*. *C. r. hedb. Séanc. Acad. Sci. Paris,* **216**, 90–1

—1944. Recherches expérimentales sur la croissance. *Bull. biol. Fr. Belg.,* **78**, 215–56

Acton, R. T., Evans, E. E., Bennett, J. C., 1969. Immunological capabilities of the oyster *Crassostrea virginica*. *Comp. Biochem. Physiol.,* **29**, 149–60

Anderson, K., and Nilsson, I., 1967. Åkersnigeln, ett växtskyddsproblem pä styva jordar. *Växtskyddsnotiser,* **31**, 67–70

Andrewartha, H. G., and Birch, L. C., 1954. *The distribution and abundance of animals.* Chicago: University of Chicago Press

Anon, 1951. *Insect pests and diseases of plants.* Govt. Printer, Brisbane, **3**, 532–3

—1953. *Land slugs and snails and their control.* U.S. Dept. of Agriculture. Farmers Bulletin no. **1895**

—1959. Landslakken. *Plantenziektenkundige Dienst Wageningen,* **74**

—1965. Sniler. Som skadedyr pä planter. *Statens Plantereras Flygeskafter,* Norway, 8/65

—1966. Snails and slugs. *New South Wales Dept., of Agriculture. Insect pest leaflet*

—1967. Slugs and snails. *U.K. Min. of Agriculture Advisory Leaflet* **115**

—1969. *Pflanzenschutzmittelverzeichnis* **1969**. Biologische Bundesanstalt für Land und Forstwirtschaft in Braunschweig. Bundesrepublik Deutschland

Arias, R. O., and Crowell, H. L., 1963. A contribution to the biology of the grey garden slug. *Bull. Calif. Acad. Sci.,* **62**, 83–97

Bailey, T. G., 1970. *Studies on organ cultures of slug reproductive tracts.* Ph.D. thesis, University of Wales

Barnes, H. F., and Weil, J. W., 1942. Baiting slugs using metaldehyde mixed with various substances. *Ann. appl. Biol.* **29**, 56–68

—1944. Slugs in gardens: their numbers, activities and distribution. Part I. *J. Anim. Ecol.*, **13**, 140–75

—1945. Slugs in gardens: their numbers, activities and distribution. Part II. *J. Anim. Ecol.*, **14**, 71–105

Barr, R. A., 1926. Some observations on the pedal gland of *Milax*. *Q. Jl. microsc. Sci.*, **70**, 647–67

—1928. Some notes on the mucous and skin glands of *Arion ater*. *Q. Jl. microsc. Sci.*, **71**, 503–25

Bayne, C. J., 1966. Observations on the composition of the layers of the egg of *Agriolimax reticulatus*, the grey field slug (Pulmonata, Stylomatophora). *Comp. Biochem. Physiol.*, **19**, 317–38

—1967. Studies on the composition of extracts of the reproductive glands of *Agriolimax reticulatus*, the grey field slug (Pulmonata, Stylomatophora). *Comp. Biochem. Physiol.*, **23**, 761–73

—1968. A study of the desiccation of egg capsules of eight gastropod species. *J. Zool. Lond.*, **155**, 401–11

Beresford-Jones, W. P., 1966. Observation on *Muellerius capillaris* (Müller) Cameron, 1927. *Res. Vet. Sci.*, **7**, 61–6, 287–91

Bett, J. A., 1960. The breeding seasons of slugs in gardens. *Proc. zool. Soc., Lond.*, **135**, 559–68

Bierbauer, J., Török, L. J., Teichmann, I., 1965. Cytologische und Histochemische Untersuchungen am Neurosekretorischen System der Augententakel von Pulmonaten. *Zool. Jb. allg. Zool. Physiol,* **71**, 545–51

Blaisdell, K., 1952. A study of the cat lungworm *Aelurostrongylus abstrusus*. Ph.D. thesis, Cornell University

Blažka, P., 1955. Temperaturadaptation des Gesamtmetabolismus bei der Weinbergschnecke *Helix pomatia*. *Zool. Jb. allg. Zool. Physiol.*, **65**, 430–8

Blezard, E., 1967. Non-marine mollusca of lakeland. *Trans. Carlisle nat. Hist. Soc.*, **11**, 48–74

Blinn, W. C., 1964. Water in the mantle cavity of land snails. *Physiol. Zool.*, **37**, 329–37

Boer, H. N., Bonga, S. E. W., and Rooyen, N. van, 1967. Light and electron microscopical investigations on the salivary glands of *Lymnaea stagnalis* I. *Z. Zellforsch. mikrosk. Anat.*, **76**, 228–47

Bonavita, A., 1967. Hygrocinèse chez la limace *Milax gagates* Draparnaud (gastéropode pulmoné). *C. r. hebd. Séanc. Acad. Sci. Paris*, **265**, 55–7

References

Bonse, H., 1934. Ein Beitrag zur Problem der Schneckenbewegung. *Zool. Jb. allg. Zool. Physiol.*, **54**, 349–84

Boonkoom, V., and Horne, F. R., 1968. A comparative study of the distribution and activities of the ornithine cycle enzymes in eight gastropods. *Amer. Zool.*, **8**, 817

Braithwaite, B. M., 1961. *Insect pests of pastures on the coast of New South Wales.* New South Wales Dept. of Agriculture

Brandenburg, J., 1966. Die Reusenformen der Cyrtocyten. Eine Beschreibung von fünf weiteren Reusengeisselzellen und eine vergleichende Betrachtung. *Zool. Beitr.*, **12**, 345–417

Brisson, P., 1968. Développement de l'embryon et de ses annexes et étude en culture *in vitro* chez les achatines (gastéropodes pulmonés). *Archs. Anat. microsc. Morph. exp.*, **57**, 345–68

Brooks, W. M., 1966. The role of the holotrichous ciliates *Tetrahymena limacina* and *Tetrahymena rostrata* as parasites of the grey field slug, *Deroceras reticulatus*. Ph.D. thesis, University of California, Berkeley

Brown, A. C., 1967. The elimination of foreign particles by the snail *Helix aspersa*. *Nature, Lond.*, **213**, 1154–5

Brown, A. C., and Brown, R. J., 1965. The fate of thorium dioxide injected into the pedal sinus of *Bullia* (Gastropoda: Prosobranchiata). *J. exp. Biol.*, **42**, 509–19

Brown, F. J., 1933. Life history of the fowl tapeworm *Davainea proglottina*. *Nature, Lond.*, **131**, 276–7

Brown, M. J., 1955. Some notes on the activities of the spotted garden slug. *J. Tenn. Acad. Sci.*, **30**, 133

Bruel, N. E. van den, and Moens, R., 1958. Remarques sur les facteurs écologiques influençant l'efficacité de la lutte contre les limaces. *Parasitica*, **14**, 135–48

Bruggen, E. F. J. van, Wiebenga, E. H., and Gruber, M., 1962. Structure and properties of hemocyanins. I. Electron micrographs of hemocyanin and apohemocyanin from *Helix pomatia* at different pH values. *J. molec. Biol.*, **4**, 1–7

Burch, J. B., 1962. *How to know the eastern land snails.* Iowa, Wm. C. Brown
—1968. Tentacle retraction in tracheopulmonata. *J. malac. soc. Aust.*, **11**, 62–7

Burton, D. W., 1962. New Zealand land slugs—Part I. *Tuatara*, **9**, 87–97
—1963a. New Zealand land slugs—Part 2. *Tuatara*, **11**, 90–6

—1963b. A revision of the New Zealand and sub-Antartic Athoracophoridae. *Trans. R. Soc. N.Z. (Zool.)*, **3**, 47–75

Burton, R. F., 1964. Variations in the volume and concentration of the blood of the snail, *Helix pomatia* L., in relation to the water content of the body. *Can. J. Zool.*, **42**, 1085–97
—1965a. Possible factors limiting the concentration of haemocyanin in the blood of the snail, *Helix pomatia* L. *Can. J. Zool.*, **43**, 433–8
—1965b. Relationships between the cation contents of slime and blood in the snail *Helix pomatia* L. *Comp. Biochem. Physiol.*, **15**, 339–45
—1968. Ionic balance in the blood of Pulmonata. *Comp. Biochem. Physiol.*, **25**, 509–16
—1969. Buffers in the blood of the snail *Helix pomatia* L. *Comp. Biochem. Physiol.*, **29**, 919–30

Campbell, J. W., and Speeg, K. V., 1968a. Arginine biosynthesis and metabolism in terrestrial snails. *Comp. Biochem. Physiol*, **25**, 3–32
—1968b. Tissue distribution of enzymes of arginine biosynthesis in terrestrial snails. *Z. vergl. Physiol.*, **61**, 164–75

Cardot, H., 1924. Observation physiologiques sur les embryons des gastéropodes pulmonés. *J. Physiol. Path. gén.*, **22**, 575–86

Carrick, R., 1938. The life history and development of *Agriolimax agrestis* L., the grey field slug. *Trans. Roy. Soc. Edinb.*, **59**, 563–97
—1942. The grey field slug *Agriolimax agrestis* L. and its environment. *Ann. appl. Biol.*, **29**, 43–55

Carriker, M. R., 1946. Morphology of the alimentary system of the snail *Lymnaea stagnalis appressa* Say. *Trans. Wis. Acad. Sci. Arts Lett.*, **38**, 1–88

Chace, L., 1953. The aerial mating of the great slug. *Discovery*, **13**, 356–9

Chapman, G., 1967. *The body fluids and their functions*. London, Edward Arnold

Cheng, T., 1964. *The biology of animal parasites*. Philadelphia, W. B. Saunders and Co.

Chétail, M., 1963. Étude de la régénération du tentacle oculaire chez un arionidae *(Arion rufus* L.) et un limacidae *(Agriolimax agrestis* L.). *Archs. Anat. micr. Morph. exp.*, **52**, 129–203

Chétail, M., and Binot, D., 1967. Particularités histochimiques de la glande et de sole pedieuses d'*Arion rufus* (Stylommatophora: Arionidae). *Malacologia*, **5**, 269–84

Clark, L. R., Geier, P. W., Hughes, R. D., and Morris, R. F., 1967. *The ecology of insect populations in theory and practice*. London, Methuen and Co.

References

Clark, R. B., and Milne, A., 1955. The sub-littoral fauna of two sandy bays on the Isle of Cumbrae, Firth of Clyde. *J. mar. biol. Ass., U.K.*, **34**, 161–80

Coleman, J. A., 1970. Studies on the statocysts of *Helix aspersa*, M.Sc. thesis, University of Wales

Collinge, W. E., 1921. The starling: Is it injurious to agriculture? *J. Minist. Agric., Fish.*, **27**, 1114–21

—1924. *The food of some British wild birds*. York

Colvin, T. A., 1962. Observations on a slug, *Vaginulus* sp. in captivity. *Proc. La. Acad. Sci.*, **25**, 122–35

Costa, G., Uhlrich, L., Kantor, F., and Holland, J. F., 1968. Production of elemental nitrogen by certain mammals including man. *Nature Lond.*, **218**, 546–51

Cottrell, G. A., and Osborne, N., 1969. A neurosecretory system terminating in the *Helix* heart. *Comp. Biochem. Physiol.*, **28**, 1455–9

Cragg, J. B. and Vincent, M. H. 1952. The action of metaldehyde on the slug *Agriolimax reticulatus* (Müller). *Ann. appl. Biol.*, **39**, 396–406

Creutz, C., 1963. Ernahrungsweiss und Aktionradius der Lachmowe. *Beitr. Vogelk.*, **9**, 3–58

Crowell, H. H., 1967. Slug and snail control with experimental poison baits. *J. econ. Ent.*, **60**, 1048–50

Crozier, W. J., and Cole, W. H., 1929. The phototropic excitation of *Limax*. *J. gen. Physiol.*, **12**, 669–74

Crozier, W. J., and Federighi, H., 1925a. Phototropic circus movements of *Limax* as affected by temperature. *J. gen. Physiol.*, **7**, 151–69

—1925b. The locomotion of *Limax*. II. Vertical ascension with added loads. *J. gen. Physiol.*, **7**, 415–19

Crozier, W. J., and Libby, R. L., 1925. Temporary abolition of phototropism in *Limax* after feeding. *J. gen. Physiol.*, **7**, 421–7

Crozier, W. J., and Wolf, E., 1928. Dark adaptation in *Agriolimax*. *J. gen. Physiol.*, **12**, 83–109

Curtis, J., 1860. *Farm insects*. London. Blackie and Son

Dainton, B. H., 1954a. The activity of slugs: I. The induction of activity by changing temperatures. *J. exp. Biol.*, **31**, 165–87

—1954b. The activity of slugs: II. The effect of light and air currents. *J. exp. Biol.*, **31**, 188–97

Dawson, R. M. C., Elliot, D. C., Elliot, W. H., Jones, K. M., 1959. *Data for biochemical research.* London, Oxford University Press

Dunnett, G. M., 1956. The Autumn and Winter mortality of starlings, *Sturnus vulgaris,* in relation to their food supply. *Ibis,* **98,** 220–30

Duthoit, C. M. C., 1961. Assessing the activity of the field slug in cereals. *Pl. Path.,* **10,** 165

—1964. Slugs and food preferences. *Pl. Path.,* **13,** 73–8

Eakin, R. M., and Brandenburger, J. L., 1967. Differentiation in the eye of a pulmonate snail *Helix aspersa. J. ultrastruct. Res.,* **18,** 391–421

—1968. Localization of vitamin A in the eye of a pulmonate snail. *Proc. natn. Acad. Sci., U.S.A.,* **60,** 140–5

Ellis, A. E., 1969. *British snails.* London, Oxford University Press

Emerson, D. N., 1967. Carbohydrate oriented metabolism of *Planorbis corneus* (Mollusca, Planorbidae) during starvation. *Comp. Biochem. Physiol.,* **22,** 571–9

Evans, W. A. L., and Jones, E. G., 1962a. Carbohydrases in the alimentary tract of the slug *Arion ater* L., *Comp. Biochem. Physiol.,* **5,** 149–60

—1962b. A note on the proteinase activity in the alimentary tract of the slug *Arion ater* L., *Comp. Biochem. Physiol.,* **5,** 223–5

Fantin, A. M. B., and Bolognari, L., 1964. Dati biochimi ed istochimici sulla mucinogenesi in *Helix pomatia.* Rc. 1st. Comp. Sci. Lett., **98B,** 343–54

Feng, S. Y., 1967. Responses of molluscs to foreign bodies, with special reference to the oyster. *Fed. Proc. Fedn. Am. Socs. exp. Biol.,* **26,** 1685–92

Fenner, T. L., 1962. *Slugs and snails.* Dept. of Agriculture, South Australia, Leaflet 3686

Filhol, J., 1938. Recherches sur la nature des lépidosomes et les phénomenes cytologiques de la secretion chez les gastéropodes pulmonés. *Arch. Anat. microsc. Morph. éxp.,* **34,** 155–439

Foster, R., 1958a. The effects of trematode metacercariae (Brachylaemidae) on the slugs *Milax sowerbii* Fér. and *Agriolimax reticulatus* Müller. *Parasitology,* **48,** 261–8

—1958b. Infestation of the slugs *Milax sowerbii* Férussac and *Agriolimax reticulatus* Müller by trematode metacercariae (Brachylaemidae). *Parasitology,* **48,** 304–11

Franc, A., 1968. Sous-classe des pulmonés. In *Traité de Zoologie. 5 Mollusques gastéropodes et scaphopodes.* Ed. Grassé, 325–607

References

Franssen, J., and Jeuniaux, C., 1965. Digestion de l'acide alginique chez les invertébrés. *Cah. Biol. mar.*, **6**, 1–21

Frömming, E., 1957. Nacktschnecken als Schädlinge in Mehlvorratskellern und über den Einfluss dieser Ernährung auf die Körperfarbe. *Z. angew. Zool.* **44**, 349–57

Gabe, M., 1965. *Neurosecretion.* Oxford, Pergamon Press

Galangau, V., 1964. Le cycle sexuel annual de *Milax gagates* Drap (gasteropodes pulmoné) et ses deux pontes. *Bull. Soc. zool. Fr.*, **89**, 510–13

Garner, J. H., 1970. A study of excretion in the slug *Agriolimax reticulatus.* Ph.D. thesis, University of Wales

Gatenby, J. B., 1918. The cytoplasmic inclusions of the germ cells part III. The spermatogenesis of some other pulmonates. *Q. Jl. miscrosc. Sci.*, **63**, 197–258

Gerhardt, U., 1934. Zur Biologie der Kopulation der Limaciden II. *Z. Morph. Ökol. Tiere.*, **28**, 229–58
—1935. Weitere Untersuchungen zur Kopulation der Nacktschneken. *Z. Morph. Ökol. Tiere.*, **30**, 297–332
—1940. Neue biologische Nacktschneckenstudien. *Z. Morph. Ökol. Tiere.*, **36**, 557–80

Germain, L., 1930. Mollusques terrestres et fluviatiles. *Faune Fr.*, **21**

Gerschenfield, H. M., 1963. Observations on the ultrastructure of synapses in some pulmonate molluscs. *Z. Zellforsch. mikrosk. Anat.*, **60**, 258–75

Getz, L. L., 1959. Notes on the ecology of slugs: *Arion circumscriptus, Deroceras reticulatum* and *D. laeve. Amer. Midl. Nat.*, **61**, 485–98

Geuze, J. J., 1968. Observations on the function and the structure of the statocysts of *Lymnaea stagnalis* (L.). *Neth. J. Zool.*, **18**, 155–204

Ghiretti, F., 1966a. Respiration. In *Physiology of the Mollusca II*, 175–208. Ed. Wilbur, K. M., and Yonge, C. M., New York, Academic Press
—1966b. Molluscan hemocyanins. In *Physiology of Mollusca II*, 233–48. Ed. Wilbur, K. M., and Yonge, C. M., New York, Academic Press

Gimingham, C. T., 1940. Some recent developments by English workers to the development of methods of insect control. *Ann. appl. Biol.* **27**, 167–8

Gimingham, C. T., and Newton, H. C. E., 1937. A poison bait for slugs. *J. Minist. Agric. Fish.*, **44**, 242–6

Goddard, C. K., and Martin, A. W., 1966. Carbohydrate metabolism. In *Physiology of Mollusca II*. 275–308. Ed. Wilbur, K. M., and Yonge, C. M.

Goddard, C. K., Nicol, P. I., and Williams, J. F., 1964. The effect of albumen gland homogenate on the blood sugar of *Helix aspersa* Müller. *Comp. Biochem. Physiol.*, **11**, 351–66

Gottfried, H., Dorfman, R. I., Forchielli, E., and Wall, P. E., 1968. Aspects of the reproductive endocrinology of the giant land slug *Agriolimax californicus*. *Gen. comp. Endocr.*, **9**, 454

Gottfried, H., Dorfman, R. I., and Wall, P. E., 1967. Steroids of invertebrates: production of oestrogens by an accessory reproductive tissue of the slug *Arion ater rufus*. *Nature, Lond.*, **215**, 409–10

Gottfried, H., and Lusis, O., 1966. Steroids of invertebrates: the *in vitro* production of 11-ketotestosterone and other steroids by the eggs of the slug *Arion ater rufus* (Linn.). *Nature, Lond.*, **212**, 1488–9

Gough, H. C., and Woods, A., 1954. Seed dressings for the control of wheat bulb fly. *Nature Lond.*, **174**, 1151–3

Gould, H. J., 1961. Observations on slug damage to winter wheat in East Anglia, 1957–1959. *Pl. Path.*, **10**, 142–6

—1962. Tests with seed dressings to control grain hollowing of winter wheat by slugs. *Pl. Path.*, **11**, 147–52

—1965. Observations on the susceptibility of maincrop potato varieties to slug damage. *Pl. Path.*, **14**, 109–11

Graetz, E., 1929. Über einige Verdaungsfermente im Kropfsafte einheimischer pulmonaten. *Zool. Jb. allg. Zool. Physiol.*, **46**, 375–412

Grassé, P. 1968. *Traité de zoologie 5* (iii) *Mollusques gastéropodes et scaphopodes*. Paris, Masson et cie.

Guyard, A., and Gomot, L., 1964. Survie et differentiation de la gonade juvenile d'*Helix aspersa* en culture organotypique. *Bull. Soc. zool. France*, **89**, 48–56

Hart, J. S., 1957. Climatic and temperature induced changes in the energetics of homeotherms. *Revue. can. Biol.*, **16**, 133–74

Hasan, S., and Vago, C., 1966. Transmission of *Alternaria brassicola* by slugs. *Pl. Dis. Reptr.* **50**, 764–7

Henderson, I. F., 1969a. A laboratory method for assessing the toxicity of stomach poisons to slugs. *Ann. appl. Biol.*, **63**, 167–71

—1969b. Laboratory methods for assessing the toxicity of contact poison to slugs. *Ann. appl. Biol.*, **62**, 363–9

Hill, R. B., and Welsh, J. H., 1966. Heart, circulation and blood cells. In *Physiology of Mollusca II*. 126–74. Ed. Wilbur, K. M., and Yonge, C. M.

References

Hirsch, J. G., and Cohn, Z. A., 1964. Digestive and autolytic functions of lysosomes in phagocytic cells. *Fedn. Proc. Fedn. AM. Socs. exp. Biol.* **23**, 1023–5

Hobmaier, A., and Hobmaier, M., 1929. Über die Entwicklung des Lungenwurmes *Sythetocaulus capillaris* in Nachtweg und Schnirkelschnecken. *Munch. tierarztl. Wschr.*, **80**, 497–500

Holtz, F., and Brand, T. von., 1940. Quantitative studies upon some blood constituents of *Helix pomatia*. *Biol. Bull.*, **79**, 423–31

Holyoak, D., 1968. A comparative study of the food of some British Corvidae. *Bird Study*, **15**, 147–53

Horne, F., and Boonkoom, V., 1970. The distribution of the ornithine cycle enzymes in twelve gastropods. *Comp. Biochem. Physiol.*, **32**, 141–53

Horstmann, H. J., 1955. Untersuchungen zur physiologie der Begattung und Befruchtung der Schlammschnecke *Lymnaea stagnalis* L. *Z. Morph. Ökol. Tiere.*, **44**, 222–68

Howes, N. H., and Wells, G. P., 1934. The water relations of slugs and snails II. Weight rhythms in *Arion ater* L. and *Limax flavus* L. *J. exp. Biol.*, **11**, 344–51

Howitt, A. J., 1961. Chemical control of slugs in orchard grass—Ladino white pastures in the Pacific Northwest. *J. econ. Ent.*, **54**, 778–81

Hunter, P. J., 1966. The distribution and abundance of slugs on an arable plot in Northumberland. *J. anim. Ecol.*, **35**, 543–7
—1967. The effect of cultivations on slugs of arable ground. *Pl. Path.*, **16**, 153–6
—1968a. Studies on slugs of arable ground II. Life cycles. *Malacologia*, **6**, 379–89
—1968b. Studies on slugs of arable ground I. Sampling methods. *Malacologia*, **6**, 369–77
—1968c. Studies on slugs of arable ground III. Feeding habits. *Malacologia*, **6**, 391–9
—1969a. Slugs and their control. *Proc. 6th British Insectic. Fungic. Conf.*, **23**, 715–20
—1969b. The extent of slug damage to wheat and potatoes in England and Wales. *N.A.A.S. Q. Rev.* **85**, 31–7

Hunter, P. J., and Johnston, D. L., 1970. Screening carbamates for toxicity against slugs. *J. econ. Ent.*, **63**, 305

Hunter, P. J., and Symonds, B. V., 1970. The distribution of bait pellets for slug control. *Ann. appl. Biol.*, **65**, 1–7

Hunter, P. J., Symonds, B. V., and Newell, P. F., 1968. Potato leaf and stem damage by slugs. *Pl. Path.*, **17**, 161–4

Hyman, L. H., 1967. *The invertebrates II. Mollusca I.* McGraw Hill Co. New York

Ikeda, K., 1937. Cytogenetic studies on the self-fertilisation of *Philomycus bilineatus*. (Studies on hermaphroditism in Pulmonata II). *J. Sci. Hiroshima Univ.*, **B5**, 66–123

Isarankura, K., and Runham, N. W., 1968. Studies on the replacement of the gastropod radula. *Malacologia*, **7**, 71–91

Jenkins, C. F. H., 1960. Slugs and snails. *Bull. Dep. Agric. West. Aust.*, **2776**

Jeuniaux, C., 1954. Sur la chitinase et la flore bacterienne intestinale des mollusques gastéropodes. *Mem. Acad. R. Belg. Cl. Sci.*, 8°, **28**, 1–45

—1963. *Chitin et chitinolyse, un chapitre de la biologie moleculaire*. Paris, Masson et cie.

Jeżewska, M. M., 1968. The presence of uric acid, xanthine and guanine in haemolymph of the snail *Helix pomatia* (Gastropoda). *Bull. Acad. pol. Sci. Cl. II Sér. Sci. biol.*, **16**, 73–6

Jeżewska, M. M., Gorzkowski, B., and Heller, J., 1963. Nitrogen compounds in snail *Helix pomatia* excretion. *Acta. biochim. pol.*, **10**, 55–65

Johnson, G., 1964. Entomology Department annual report. *Rep. Rothamsted. exp. Sta.*, 146–60

Johnson, G., 1965. Entomology Department annual report. *Rep. Rothamsted. exp. Sta.*, 177–94

Johnson, G., 1967. Entomology Department Annual report. *Rep. Rothamsted. exp. Sta. 1966*, 197–8

Jones, H., 1968. Aspects of the physiology of *Patella vulgata*. Ph.D. Thesis, University of Hull

Joosse, J., Boer, M. H., and Cornelisse, C. J., 1968. Gametogenesis and oviposition in *Lymnaea stagnalis* as influenced by γ-irradiation and hunger. *Symp. zool. Soc. Lond.*, **22**, 213–35

Joosse, J., and Geraerts, W. J., 1970. On the influence of the dorsal bodies and the adjacent neurosecretory cells on the reproduction and metabolism of *Lymnaea stagnalis*. *Gen. comp. Endocrinol.*, **13**, 511

Jourdain, S., 1879. On the termination of the visceral arterioles of *Arion rufus*. *Ann. nat. Hist.*, **3**, 243

References

Kalmus, H., 1942. Anemotaxis in soft-skinned animals. *Nature, Lond.*, **150**, 524

Kerkut, G. A., 1959. A neurophysiological analysis of behaviour. *15th Int. Congr. Zool. Proc.*, 845–7

Kerkut, G. A., and Cottrell, G. A., 1962. Amino acids in the blood and nervous system of *Helix aspersa*. *Comp. Biochem. Physiol.*, **5**, 227–30

Kerkut, G. A., and Laverack, M. S., 1957. The respiration of *Helix pomatia*, a balance sheet. *J. exp. Biol.*, **34**, 97–105

Kerth, K., and Krause, G., 1969. Untersuchungen mittels Röntgenbestrahlung über den Radula-Ersatz der Nacktschnecke *Limax flavus* L. *Wilhelm Roux' Archiv.*, **164**, 48–82

Kieckebusch, W., 1953. Beitrag zur Physiologie des chemischen Sinnes von *Helix pomatia*. *Zool. Jb. allg. Zool. Physiol.*, **64**, 153–82

Kingston, N., and Freeman, R. S., 1959. On the trematodes *Brachylecithum arfi* sp. nov. (Dicrocoeliidae) and *Tanaisia* sp (Encotylidae) from the ruffled grouse, *Bonasa umbellus* L. *Can. J. Zool.*, **37**, 121–7

Kittel, R., 1956. Untersuchungen über den Geruchs—und Geschmackssin bei den Gattungen *Arion* und *Limax* (Mollusca: Pulmonata). *Zool. Anz.*, **157**, 185–95

Koshman, R. W. and Serra, J. A., 1967. Spireme karyodieresis: a new type of reductional division. *Can. J. Genet. Cytol.*, **9**, 31–7

Krijgsman, B. J., and Divaris, G. A., 1955. Contractile and pacemaker mechanisms of the heart of molluscs. *Biol. Rev.*, **30**, 1–39

Kugler, O. E., 1965. A morphological and histochemical study of the reproductive system of the slug *Philomycus carolinianus* (Bosc). *J. Morph.*, **116**, 117–32

Kuhlmann, D., 1966. Der Dorsalkorper der Stylommatophoren (Gastropoda). *Z. wiss. Zool.*, **173**, 218–31

—1969. Bestimmung des DNS-gehaltes in Zellkernen des Nervengewebes von *Helix pomatia* und *Planorbarius corneus*. *Experientia.*, **25**, 848–9

Lane, N. J. 1962. Neurosecretory cells in the optic tentacles of certain pulmonates. *Q. Jl. Microsc. Sci.*, **103**, 211–26

Langlois, T. H., 1963. The conjugal behaviour of the introduced European giant garden slug *Limax maximus* as observed on S. Bass Island, Lake Erie. *Ohio. J. Sci.*, **65**, 208–304

Lanza, B., and Quattrini, D., 1964. Richerche sulla biologia dei Veronicellidae (Gastropoda Soleolifera). I. la riproduzione in isolamento individuale di *Vaginulus borellianus* (Colosi) e di *Laevicaulis alte* (Férussac). *Monit. zool. ital.*, **72**, 93–141

Lapage, G., 1956. *Veterinary parasitology*. London, Oliver and Boyd

Laryea, A. A. 1969. The Arterial Gland of *Agriolimax Reticulatus* (Pulmonata, Limacidae) *Malacologia*, **9**, 273
—1970. Studies on the nervous system of the slug *Agriolimax reticulatus*. Ph.D. thesis, University of Wales

Laverack, M. S., 1968. On superficial receptors. *Symp. zool. Soc. Lond.*, **13**, 299–326

Laviolette, P., 1950a. Sur un retard de la maturité genitale observe chez *Arion rufus*. *Bull bimens. Soc. linn. Lyon.*, **19**, 52–6
—1950b. Rôle de la gonade dans la morphogénèse du tractus génital chez quelques mollusques limacidae et arionidae. *C. r. Hebd. Seanc. Acad. Paris.*, **213**, 1567–9
—1953. Regeneration germinale après castration chez un mollusque gastropode pulmoné *Arion rufus*. *14th Int. Congr. Zool.* 233–5
—1954. Rôle de la gonade dans le déterminisme glandulaire du tractus génital chez quelques gastéropodes arionidae et limacidae. *Bull. biol. Fr. Belg.*, **88**, 310–32

Laviolette, P., and Cuir, P., 1959. Actions des rayons x sur la gonade d'*Arion rufus* L. (mollusque gasteropode). *Archs. Anat. micr. Morph. exp.*, **48**, 25–47

Laviolette, P., and Voulot, C., 1966. Étude de la croissance des mollusques irradiés *(Arion rufus* et *Limnaea stagnalis)*. *Bull. biol. Fr. Belg.*, **95**, 679–94

Lawrence, A. L., and Lawrence, D. C., 1967. Sugar absorption in the intestine of the chiton *Cryptochiton stelleri*. *Comp. Biochem. Physiol.*, **22**, 341–57

Lee, T. W., and Campbell, J. W., 1965. Uric acid synthesis in the terrestrial snail *Otala lactea*. *Comp. Biochem. Physiol.*, **15**, 457–68

Lewis, R. D., 1967. Studies on the activity of the slug *Arion ater* (L.). Ph.D. thesis, University of Wales
—1969a. Studies on the locomotor activity of the slug *Arion ater* (Linnaeus) I. Humidity, temperature and light reactions. *Malacologia*, **7**, 295–306
—1969b. Studies on the locomotor activity of the slug *Arion ater* (Linnaeus). II. Locomotor activity rhythms. *Malacologia*. **7**, 307–12

Liang, C., and Rosenberg, H., 1968. On the distribution and biosynthesis of 2-aminoethyl phosphonate in two terrestrial molluscs. *Comp. Biochem. Physiol.*, **25**, 673–81

References

Likharev, I. M., and Rammel'meier, E. S., 1962. *Terrestrial molluscs of the fauna of the U.S.S.R.* Jerusalem, Israeli Program for Scientific Translations

Lissmann, H. W., 1945a. The mechanism of locomotion in gastropod molluscs. I. Kinematics. *J. exp. Biol.*, **21**, 58–69

—1945b. The mechanism of locomotion in gastropod molluscs. II. Kinetics. *J. exp. Biol.*, **22**, 37–50

Lloyd, D., 1969. Studies on defence mechanisms in *Oxychilus* spp. (Fitzinger) (Mollusca, Pulmonata, Zonitidae). Ph.D. thesis, University of Wales

Lockie, J. D., 1956. The food and feeding behaviour of the Jackdaw, Rook and Carrion Crow. *J. anim. Ecol.*, **25**, 421–8

Lontie, R., Brauns, G., Cooreman, H., and Vauclef, A., 1962. Distribution of α- and β-haemocyanins in the blood of *Helix pomatia. Archs. Biochem. Biophys.* Suppl. 1., 295–300

Lusis, O., 1961. Post embryonic changes in the reproductive system of the slug *Arion rufus* L. *Proc. zool. Soc. Lond.*, **137**, 433–68

—1966. Changes induced in the reproductive system of *Arion ater rufus* L. by varying environmental conditions. *Proc. malac. Soc. Lond.*, **37**, 19–26

Luther, A., 1915. Zuchtversuche an Ackerschnecken *(Agriolimax reticulatus* Müll und *Agr. agrestis* L*). Acta Soc. Fauna Flora fenn.*, **40**, 1–42

Machin, J., 1964. The evaporation of water from *Helix aspersa* I. The nature of the evaporating surface. *J. exp. Biol.*, **41**, 759–69

—1965. Cutaneous regulation of evaporative water loss in the common garden snail *Helix aspersa. Naturwissenschaften.*, **52**, 18

Malek, E. A., 1962. *Laboratory guide and notes for medical malacology.* Minneapolis, Burgess Publishing Co

Märkel, K., 1957. Bau und Funktion der Pulmonatenradula. *Z. wiss. Zool.*, **160**, 213–89

Martin, A. W., Harrison, F. M., Huston, M. J., and Stewart, D. M., 1958. The blood volumes of some representative molluscs. *J. exp. Biol.*, **35**, 260–79

Martin, A. W., Stewart, D. M., and Harrison, F. M., 1965. Urine formation in a pulmonate land snail *Achatina fulica. J. exp. Biol.*, **42**, 99–123

Martoja, M., 1964a. Contribution a l'étude de l'appareil digestif et de la digestion chez les gastéropodes carnivores de la famille nassaridés. *Cellule*, **64**, 237–334

—1964b. Dévelopment de l'appareil reproducteur chez les gastéropodes pulmonés. *Annee biol.*, **3**, 199–232

Maury, M. F., and Reygrobellet, D., 1963. Sur les distinctions specifiques chez les mollusques limacidés du genre *Deroceras*. *C. r. hebd. Séanc. Acad. Sci. Paris*, **288**, 2902–3

Meenakshi, V. R., and Scheer, B. T., 1968. Studies on the carbohydrates of the slug *Ariolimax columbianus* with special reference to their distribution in the reproductive system. *Comp. Biochem. Physiol.*, **26**, 1091–7

Michelson, E. H., 1964. Miracidia-immobilizing substances in extract prepared from snails infected with *Schistosoma mansoni*. *Am. J. trop. Med. Hyg.*, **13**, 36–42

Milne, A., 1957. Theories of natural control of insect populations. *Cold Spring Harb. Symp. quant. Biol.*, **22**, 253–67
—1962. On a theory of natural control of insect populations. *J. theor. Biol.*, **3**, 19–50

Minker, E., and Koltai, M., 1961. Untersuchungen an isolierten Gastropodenorganen. *Acta. biol. Hung.*, **12**, 199–209

Moens, R., François, E., and Riga A., 1966. Les radioisotopes en écologie animale; déplacements dans un milieu naturel de *Agriolimax reticulatus* O.F. Müller. *Meded. Rijksfac.*, **31**, 1032–42

Mol, J.-J. van., 1961. Étude histologique de la gland cephalique au cours de la croissance chez *Arion rufus* Linne. *Annls. Soc. r. zool. Belg.*, **91**, 45–56
—1967. Étude morphologique et phylogénétique du ganglion cérébroïde des gastéropode pulmonés (mollusques). *Mem. Acad. r. Belg. Cl. Sci.*, 8°, **37**, 1–168

Morris, R. F., 1959. Single factor analysis in population dynamics. *Ecology*, **40**, 580–8
—1963. Predictive population equations based on key factors. *Mem. ent. Soc. Can.*, **32**, 16–21

Morton, J. E., 1958. *Molluscs*. London, Hutchinson University Library

Müller, G., 1956. Morphologie Lebensblauf und Bildungsort der Blutzellen von *Lymnaea stagnalis* L. *Z. Zellforsch. mikrosk. Anat.*, **44**, 519–56

Newell, P. F., 1965. The behaviour and distribution of slugs. Ph.D. Thesis, University of London
—1966. The nocturnal behaviour of slugs. *Med. biol. Illust.*, **16**, 146–59
—1968. The measurement of light and temperature as factors controlling the surface activity of the slug *Agriolimax reticulatus* (Müller). *Symp. Brit. ecol. Soc.*, **8**, 141–6

References

Newell, P. F., and Newell, G. E., 1968. The eye of the slug *Agriolimax reticulatus* (Müll). *Symp. zool. Soc. Lond.*, **23**, 97–111

Nicholson, A. J., 1954. An outline of the dynamics of animal populations. *Aust. J. Zool.*, **2**, 9–65

—1957. The self-adjustment of populations to change. *Cold Spring Harb. Symp. quant. Biol.*, **22**, 153–72

—1958. Dynamics of insect populations. *A. Rev. Ent.*, **3**, 107–36

Nielsen, C. O., 1962. Carbohydrases in soil and litter invertebrates. *Oikos*, **13**, 200–15

—1963. Laminarinases in soil and litter invertebrates. *Nature, Lond.*, **199**, 1001

Nisbet, R. H., and Plummer, J. M., 1966. Further studies on the fine structure of the heart of Achatinidae. *Proc. malac. Soc. Lond.*, **37**, 199–208

—1968. The fine structure of cardiac and other molluscan muscle. *Symp. zool. Soc. Lond.*, **22**, 193–211

Nolte, A., and Machemer-Röhnisch. S., 1966. Experimentelle Untersuchungen zur Funktion der Dorsalkörper bei *Lymnaea stagnalis* L. (Gastropoda, Basommatophora). *Z. wiss. Zool.*, **73**, 232–44

Nopp, H., and Farahat, A. Z., 1967. Temperatur und Zellstoffwechsel bei Heliciden. *Z. Vergl. Physiol.*, **55**, 103–18

Oldham, C., 1942. Auto-fecondation and duration of life in *Limax cinereo-niger*. *Proc. malac. Soc. Lond.*, **25**, 9–10

Owen, G., 1966. Digestion. In *Physiology of Mollusca*, II, 53–96. Ed. Wilbur, K. M., and Yonge, C. M.

Pallant, D., 1969. The food of the grey field slug (*Agriolimax reticulatus*) in woodland. *J. anim. Ecol.*, **38**, 391–8

Pelluet, D., 1964. On the hormonal control of cell differentiation in the ovotestis of slugs (Gasteropoda; Pulmonata). *Can. J. Zool.*, **42**, 195–9

Pelluet, D., and Henderson, N. E., 1954. The effects of visible radiation on the ovotestis of the slug *Deroceras reticulatum*. *Proc. Nova. Scotian Inst. Sci.*, **23**, 420–1

Pelluet, D., and Lane, N. J., 1961. The relation between neurosecretion and cell differentiation in the ovotestis of slugs (Gasteropoda: Pulmonata). *Can. J. Zool.*, **39**, 789–805

Pelseneer, P., 1935. *Essai d'ethologie zoologique*. Brussels, Acad. Roy. Belg. Classe. Sci. Publ. Fond. Agathon Potter

Peyer, B., 1954. Sull 'accoppiamento nel genus *Limax*. *Boll. Soc. ticin. Sci. Nat.*, **49**, 82–7

Pilsbry, H. A., 1948. Land mollusca of North America (north of Mexico): Vol II, part II. *Acad. nat. Sci. Philad.*

Pinder, L. C. V., 1969. The biology and behaviour of some slugs of economic importance. *Agriolimax reticulatus, Arion hortensis* and *Milax budapestensis*. Ph.D. thesis, University of Newcastle-upon-Tyne

Pohunkova, H., 1967. The ultrastructure of the lung of the snail *Helix pomata*. *Folia morph.*, **15**, 250–7

Prokopič, J., and Ždarská, Z., 1958. *Arion rufus* (L) and *Vitrina pellucida* (Müller) as intermediate hosts of the tapeworm *Anomotaenia subterranea*. Cholodkowsky, 1906. *Mem. Soc. Zool. Tchécoslov.*, **22**, 1–5

Prosser, C. L., and Brown, F. A., 1961. *Comparative animal physiology*. Philadelphia, Saunders and Co.

Quattrini, D., and Lanza, B., 1965. Ricerche sulla biologia dei Veronicellidae (Gastropoda Soleolifera). II. Struttura della gonade ovogenesi e spermatogenesi in *Vaginulus borellianus* (Colosi) e in *Laevicaulis alte* (Férussac). *Monitore zool. ital.*, **73**, 3–60

Quick, H. E., 1947. *Arion ater* (L) and *A. rufus* (L.) in Britain and their specific differences. *J. Conch, Lond.*, **22**, 249–61
—1949. Synopses of the British fauna. No. 8. Slugs (Mollusca) (*Testacellidae, Arionidae, Limacidae*). London, Linnean Society
—1960. British slugs (Pulmonata: Testacellidae, Arionidae, Limacidae). *Bull. Br. Mus. nat. Hist. Zool.*, **6**, no. 3, 103–226

Raven, C. L., 1958. *Morphogenesis: The analysis of molluscan development*. London, Pergamon Press

Razet, P., and Dagobert, C., 1968. Recherches des enzymes de la chain de l'uricolyse chez les mollusques terrestres et dulçaquicoles. *Archs. Sci. physiol.*, **22**, 173–81

Regondaud, J., 1964. Origine embryonnaire de la cavité pulmonaire de *Lymnaea stagnalis* L. Considérations particulières sur la morphogenèse de la commissure viscérale. *Bull. biol. Fr. Belg.*, **48**, 433–71

Renzoni, A., 1968. Osservazioni istologiche, istochimiche ed ultrastrutturali sui tentacoli di *Vaginulus borellianus* (Colosi) Gastropoda Soleolifera. *Z. Zellforsch. mikrosk., Anat.* **87**, 350–76
—1969. Observations on the tentacles of *Vaginulus borellianus* Colosi (Mollusca, Gastropoda, Soleolifera). *Veliger*, **12**, 176–81

Richards, O. W., and Southwood, T. R. C., 1968. The abundance of insects. Introduction. In *Insect abundance* Ed. Southwood, T. R. C., Oxford, Blackwell

References

Richter, E., 1935. Der Bau der Zwitterdrüse und die Entstehung der Geschlechtszellen bei *Agriolimax agrestis*. *Z. Naturw.*, **69**. 507–44

Rising, T. L., and Armitage, K. B., 1969. Acclimation to temperature by the terrestrial gastropods, *Limax maximus* and *Philomycus carolinianus*: oxygen consumption and temperature preference. *Comp. Biochem. Physiol.*, **30**, 1091–1114

Roach, D. K., 1963. Analysis of the haemolymph of *Arion ater* L. (Gastropoda: Pulmonata). *J. exp. Biol.*, **40**, 613–23
—1966. Studies of some aspects of the physiology of *Arion ater*. Ph.D. thesis, University of Wales
—1968. Rhythmic muscular activity in the alimentary tract of *Arion ater* (L) (Gastropoda: Pulmonata). *Comp. Biochem. Physiol.*, **24**, 865–78

Rogers, D. C., 1969. The fine structure of the collar cells in the optic tentacles of *Helix aspersa*. *Z. Zellforsch. mikrosk. Anat.*, **102**, 113–28

Röhlich, P. and Bierbauer, J., 1966. Electron microscopic observations on the special cells of the optic tentacle of *Helicella obvia* (Pulmonata). *Acta biol. Hung.*, **17**, 359–73

Rosén, 1952. The problem of phagocytosis in *Helix pomatia* L. *Ark. Zool.*, **3**, 33–50

Rosenbaum, R. M., and Ditzion, B., 1963. Enzymic histochemistry of granular components in digestive gland cells of the Roman snail *Helix pomatia*. *Biol. Bull.*, **124**, 211–24

Roy, A., 1963. Étude de l'acclimation thermique chez la limace *Arion circumscriptus*. *Can. J. Zool.*, **41**, 671–98

Runham, N. W., 1962. Rate of replacement of the molluscan radula. *Nature, Lond.*, **194**, 992–3
—1963. A study of the replacement mechanism of the pulmonate radula. *Q. Jl. microsc. Sci.*, **104**, 271–7
—1969. The use of the scanning electron microscope in the study of the gastropod radula: the radulae of *Agriolimax reticulatus* and *Nucella lapillus*. *Malacologia*, **9**, 179–85

Runham, N. W., and Laryea, A. A., 1968. Studies on the maturation of the reproductive system of *Agriolimax reticulatus* (Pulmonata: Limacidae). *Malacologia*, **7**, 93–108

Runham, N. W., and Thornton, P. R., 1967. Mechanical wear of the gastropod radula: a scanning electron microscope study. *J. Zool.*, **153**, 445–52

Ruppel, R. F., 1959. Effectiveness of sevin against the grey garden slug. *J. econ. Ent.*, **52**, 360

Salt, G., and Hollick, F. S. J., 1944. Studies of wireworm populations I. A census of wireworms in pastures. *Ann. appl. Biol.*, **31**, 52

Schuurmans-Stekhoven, J. H., 1920. Über die Atmung der Schnecken *Limax agrestis* und *Helix pomatia*. *Tijdschr. ned. dierk. Vereen.*, **18**, 1–43

Schwalbach, G., Lickfield, K. G., Hahn, M., 1963. Der mikromorphologische Aufbau des Linsenauges der Weinbergschnecke (*Helix pomatia* L.) *Protoplasma.*, **56**, 242–73

Schwartzkopff, J., 1954. Über die Leistung des Isolierten Herzens der Weinbergschnecke *(Helix pomatia)* im Künstlichen Kreislauf. *Z. vergl. Physiol.*, **36**, 543–94
—1956. Herzfrequenz und Körpergrosse bei Mollusken. *Zool. Jb. Suppl.*, **20**, 463–9

Segal, E., 1959. Studies on annual rhythms of egg laying in slugs under constant conditions. *Anat. Rec.*, **134**, 636

Serra, J. A., and Koshman, R. W., 1967. Nuclear changes of treptional nature in the differentiation of sperm nurse cells in snails. *Can. J. Genet. Cytol.*, **9**, 23–30

Simpson, L., Bern, H. A., and Nishioka, R. S., 1966. Examination of the evidence for neurosecretion in the nervous system of *Helisoma tenue* (Gastropoda Pulmonata). *Gen. comp. End.*, **7**, 525–48

Smith, B. J., 1966. Maturation of the reproductive tract of *Arion ater* (Pulmonata: Arionidae). *Malacologia*, **4**, 325–49

Smith, R., 1967. Recent developments in integrated control. *Proc. 4th Brit. Insectic. Fungic. Conf.*, 464–71

South, A., 1964. Estimation of slug populations. *Ann. appl. Biol.*, **53**, 251–9
—1965. Biology and ecology of *Agriolimax reticulatus* (Müll) and other slugs: spatial distribution. *J. anim. Ecol.*, **34**, 403–19

Southwood, T. R. E., 1966. *Ecological methods with particular reference to the study of insect populations.* London, Methuen and Co., Ltd.

Speeg, K. V., and Campbell, J. W. 1968a. Purine biosynthesis and excretion in *Otala* (=*Helix*) *lactea:* an evaluation of the nitrogen excretory potential. *Comp. Biochem. Physiol.*, **26**, 579–95
—1968b. Formation and volatilization of ammonia gas by terrestrial snails. *Am. J. Physiol.*, **214**, 1392–1402

Spoek, G. L., Bakker, H., and Wolvekamp, 1964. Experiments on the haemocyanin-oxygen equilibrium of the blood of the edible snail (*Helix pomatia* L). *Comp. Biochem. Physiol.*, **12**, 209–21

Stephenson, J. W., and Knutson, L. V., 1966. A résumé of recent studies of invertebrates associated with slugs. *J. econ. Ent.,* **59,** 356–60

Stern, G., 1969. Bilan energetique de la limace *Arion rufus* (Mollusque Pulmoné) en période de croissance. *C. r. hebd. Séanc. Acad. Sci. Paris,* **269,** 1015–18

Stone, B. A., and Morton, J. E., 1958. The distribution of cellulases and related enzymes in mollusca. *Proc. malac. Soc. London.,* **33,** 127–41

Strasdine, G. A., and Whitaker, D. R., 1963. On the origin of the cellulase and chitinase of *Helix pomatia. Can. J. Biochem. Physiol.,* **41,** 1621–6

Strickland, A. H., 1965. Pest control and productivity in British agriculture. *Jl. R. Soc. Arts.,* **113,** 62–81

Stringer, A., 1946. Action of metaldehyde on slugs. *Rep. agric. hort. Res. Stn. Univ. Bristol,* 87–8

Taylor, J. W., 1907. *Monograph of the land and freshwater mollusca of the British Isles.* (*Testacellidae, Limacidae, Arionidae*). Leeds, Taylor Brothers

Thomas, D. C., 1944. Field sampling for slugs. *Ann. appl. Biol.,* **31,** 163–4
—1948. The use of metaldehyde against slugs. *Ann. appl. Biol.,* **35,** 207–27

Thompson, G. A., and Hanahan, D. J., 1963. Identification of α glyceryl ether phospholipids as major lipid constituents in two species of terrestrial slug. *J. biol. Chem.,* **238,** 2628–31

Thorpe, D. L., and Hope, L. E., 1908. Report on the food of the black-headed gull. *Cumberland County Council, 1907,* p. 3–27

Turk, F. A., and Phillips, S. M., 1946. A monograph of the slug mite *Riccardoella limacum* (Schrank). *Proc. zool. Soc. Lond.,* **115,** 448–71

Turner, R. S., 1966. Patterns of form and function in the central nervous system of *Ariolimax. J. comp. Neurol.,* **128,** 51–62

Varley, G. C., and Gradwell, G. R., 1960. Key factors in population studies. *J. anim. Ecol.,* **29,** 399–401

Verdcourt, B., 1947. The speed of land mollusca. *J. Conch, Lond.,* **22,** 269–70

Vernon, J. D. R., 1970a. A preliminary study of the food of the common gull on grassland in the autumn and winter. *Bird study,* **17**
—1970b. On the feeding habitats and food of the black-headed and common gull. (In press)

Vlieger, T. A. de, 1968. An experimental study of the tactile system of *Lymnaea stagnalis* (L). *Neth. J. Zool.,* **18.** 105–54

Voogt, P. A., 1967. Biosynthesis of 3β-sterols in a snail *Arion rufus* L. from $1-^{14}$C-acetate. *Archs int. Physiol. Biochim.*, **75**, 492–500

Vorwohl, G., 1961. Zur Funktion der Exkretionsorgane von *Helix pomatia* L. und *Archachatina ventricosa* Gould. *Z. vergl. Physiol.*, **45**, 12–49

Walker, G., 1969. Studies on digestion of the slug *Agriolimax reticulatus* (Müller). (Mollusca, Pulmonata, Limacidae). Ph.D. thesis University of Wales

Watson, H., 1943. Notes on a list of the British non-marine mollusca. *J. Conch. Lond.*, **22**, 13–72

Webb, G. R., 1961. The sexology of three species of limacid slugs. *Gastropodia*, **1**, 31–44

Webb, W. M., 1893. On the manner of feeding in *Testacella scutulum*. *Zoologist, 3rd. Ser.*, **17**, 281–9

Webley, D., 1962. Experiments with slug baits in South Wales. *Ann. app Biol.*, **50**, 129–36
—1964. Slug activity in relation to weather. *Ann. appl. Biol.*, **53**, 407–14
—1965. Aspects of trapping slugs with metaldehyde and bran. *Ann. appl. Biol.*, **56**, 37–45

Weiss, M. 1968. Zur embryonalen und postembryonalen Entwicklung des Mitteldarmes bei Limaciden und Arioniden (Gastropoda, Pulmonata). *Revue Suisse Zool.*, **75**, 157–225

Wester, R. E., Goth, R. W., and Webb, R. E., 1964. Transmission of downy mildew (*Phytopthera phascoli*) of lima beans by slugs. *Phytopathology*, **54**, 749

White, A. R., 1959. Infestation of slug mite (*Riccardoella limacum*). (Schranck) on various species of slugs. *Entomologist's mon. Mag.*, **95**, 14

Wilkinson, A. T. S., 1964. Control of slugs. *Canada Dept. of Agriculture*, **1213**

Williams, D. W., 1942. Studies on the biology of the larva of the nematode lungworm *Muellarius capillaris* in molluscs. *J. anim. Ecol.*, **11**, 1–8

Williamson, M. H., 1959. Studies on the colour and genetics of the black slug. *Proc. R. Soc. Edinb.*, **27**, 87–93

Wilson, F., 1968. Insect abundance: prospect. In *Insect Abundance* Ed. Southwood, T. R. C., 143–58. Oxford, Blackwell

Winfield, A. L., Wardlow, L. R., and Smith, B. F., 1967. Further observations on the susceptibility of potato cultivars to slug damage. *Pl. Path.*, **16**, 136–8

References

Wolf, E., 1927. Geotropism of *Agriolimax*. *J. gen. Physiol.*, **10,** 757–65

Wolff, H. G., 1968. Elektrische Antworten der Statonerven der Schnecken (*Arion empiricorum,* und *Helix pomatia*) auf Drehreizung. *Experientia,* **24,** 848–9

—1969. Einige Ergebnisse zur Ultrastruktur der Statocysten von *Limax maximus, Limax flavus,* und *Arion empiricorum* (Pulmonata). *Z. Zellforsch. mikrosk. Anat.*, **100,** 251–70

Wondrak, G., 1968. Elektronenoptische Untersuchungen der Körperdecke von *Arion rufus* L. (Pulmonata). *Protoplasma,* **66,** 151–71

—1969. Elektronenoptische Untersuchungen der Drüsen- und Pigmentzellen aus der Korperdecke vor *Arion rufus* (L) (Pulmonata) *Z. mikrosk. anat. Forsch.*, **80,** 17–40

Wood, E. J., Salisbury, C. M., and Bannister, W. H., 1968. The binding of copper to haemocyanin. *Biochem. J.*, **108,** 26–7P

Yonge, C. M., 1937. Evolution and adaptation in the digestive system of the metazoa. *Biol. Rev.*, **12,** 87–115

Zs-Nagy, I., and Sakharov, D. A., 1969. Axo-somatic synapses in procerebrum of gastropodia. *Experientia.*, **25,** 258–9

INDEX

Italic figures refer to illustrations and **bold figures** to main references

ACCESSORY RETINA, **102**, *103*, 104
Acetyl-choline, 68, 112
Achatina, 81
Acid phosphatase, 55
Active transport, 57
Activity, 71, 72, 104, 107, **108–9**, 123, 129, 132, 140, 145, 146, 149, 150, 154
Aelurostrongylus abstrusus, 137, 142
Aggregation, **126**, 127
Agriolimax, 12, *13*, 14, 15, 19, **30–1**, 76, 77, 84, 90, 93, 97, 107, 108, 116, 135, 137
Agriolimax agrestis, 21, **32**, 48–9, 56, 64, 65, 67, 91, 97, 102, 136–7, 141
Agriolimax caruanae, 14, **32–3**, 44, 126, 136
Agriolimax laevis, **32**, 91, 97, 104, 107, 125, 129, 136–7
Agriolimax meridionalis, 91, 97
Agriolimax reticulatus, 12, *14*, *18*, 21, Fig. 13, **31–2**, *78*
  arterial gland, **70**, 114
  behaviour, 73, 104, 105–6, 108–9
  blood, 73, 74, 75
  blood system, **65–8**
  control, 143, 144, 146, 147, 149–50 153
  development, 94–6
  diet, 37–8
  digestive system, *16*, *45*, **46–58**
  ecology, 116, 118, 119, 120, 121, 123–9, 133, 134, 145
  excretory system, 69, **77–82**
  feeding tracks, Fig. 24
  growth, 96
  jaw, Fig. 23
  mucus, 101

  nervous system, 109–11, Fig. 44
  parasites, 136–7, 141–2
  radula, Fig. 22, *42*, 44
  reproductive system, *19*, **83–94**, 96, 114–15, Fig. 41, Fig. 43
  respiration, 63
  tentacles, 102–6
  water relations, 71–2
Albumen, 19, **85**, 92, 94, 95
Albumen gland, 17, *19*, 85, *86*, 92, 114
Alginase, 49
Allogona profunda, 71
Ammonia, 80–1
Amoebocytes, 56, 68, **69–70**
Amylase, 46, 48, 50
Anadenulus, 27
Anadenulus cockerelli, 27
Anadenus, 27
Anadininae, 27
Anafilaroides rostratus, 137, 142
Anatomy, internal, 15–20
  external, 12–15
Aneita, *13*
Aneiteinae, 35
Angiostrongylus cantonensis, 137, 142
Anomotaenia subterranea, 136, 142
Antibodies, 69
Aorta, 17, 67
Aphids, 139
Archachatina, 81
Arginine, 61, 81
Ariolimacinae, 27
Ariolimax, 27, 109
Ariolimax carolinianus, 115
Ariolimax columbianus, 60, 61, 115, 136, 137
Arion, 10, 11, 19, 24, 65, 80, 100, 101, 135, 136, 137

Arion ater, 13, **24–5**, Fig. 12
  behaviour, 104, 105, 107, 108
  blood system, 67, 71, 73, 74, 75
  breathing, 65
  diet, 38
  digestive enzymes, 46, 48–9
  digestive system, 40, 47, 50, 52, 54
  ecology, 116, 126
  foot, 15, 101
  growth, 38, 96
  metabolism, 61
  nervous system, 113
  parasites, 136
  radula, 44
  reproduction, 24, 83, 85, 88, 91, 92, 93
  skin, 14
  statocysts, 106, 107
  tentacles, 102
  water relations, 71, 72
Arion ater ater, 24–5, 74, 97, 136, 137
Arion ater rufus, 25, 37, 59, 74, *90*, 96, 136
Arion circumscriptus, 26
Arion fasciatus, 24, **26**, 48–9, 63, 64, *90*, 96, 126, 129, 136, 137
Arion hortensis, **25–6**, Fig. 11
  control, 143, 144, 145, 153
  ecology, 116, 118, 120, 121, 123, 124, 132, 133
  growth, 96
  locomotion, 100
  parasites, 136–7
  reproduction, 24
  respiration, 63
  skin, 15
Arion intermedius, 96, 136–7
Arion lusitanicus, 24
Arion silvaticus, 26
Arion subfuscus, 24, 71, 96, 101, 136–7
Arionidae, 12, 22, **24–7**, 95, 114
Arioninae, 24
Arterial gland, **70**, 77, 113–14
Arteries, 17, *66*, 68, 70, 110
Arterioles, 17, 64, 68
Assimilation, 59
Athoracophoridae, 13, 14, 15, 16, 17, 22, 23, **34–5**
Athoracophorinae, 35
Athoracophorus, 35
Atopos, 36
Aulacopod, 15
Auricle, 17, 67, 77
Auriculo-ventricular junction, 67, 68
Australorbis, 70

BACTERIA, 47, 50, 52, 69, 80
Bait, 120, 147, 149, 152, 153, 154
  attractiveness, 147, 149
Barley, 140
Basommatophora, 11, 22
Beans, 125, 142
Behaviour, 73, 91, 94, 104, 108, 151
Bicarbonate, 61, 74, 75, 76
Binneya, 24, 27
Biological control, 143
Birds, 134, 135, 143, 152, 154
Bladder, 77
Blastula, 92, 94
Blood, 59–60, 65, 67, 68, 69, 71, 75–7, 80, 81, 82, 99, 101, 108
  amino acids, 61
  buffering, 75
  carbohydrate, 59–60
  composition, 73–7
  pH, 75, 76, 81
  pressure, 68, 72, 82, 87, 99, 101
  protein, 74
  sinuses, 17, 66
  system, 17, **65–71**, 77
  vessel anastomoses, 64, 68
  volume, 71, 73
Blowing, 141
Body shape, 73
Body wall, 20, 61, 65, 66, 101
Boettgerilla, 29
Bohr effect, 76
Brachycoelum obesum, 137, 142
Brachylaema, 136, 141
Brachylaema nicolli, 136, 142
Brachylaema virginianum, 136, 142
Brachylecithum orfi, 141
Brain (see also Ganglia), 63, 112, 115
Breathing, 64–5
Breeding season, 83, 84, 93, 96, 126, 132, 153
Brussels sprouts, 140–1
Buccal cavity, 15, 39, *40*, 41, 57, 147
Buccal mass, 20, 39, 40
Bulbs, 140
Bullia, 69

CABBAGE, 143
Cabbage leaf spot, 142
Calcium, 11, 12, 70, 73, 74, 85, 93, 94, 101, 113
Calcium cells, 52, *53*, 55
Capillaries, 17, 65
Carbamates, 146, 147, 148, 149, 150, 154
Carbaryl, 147
Carbohydrases, 47, 48, 56

Index

Carbohydrate metabolism, **59–60**, 64
Carbon dioxide, 62, 64, 65, 75, 76, 77
Carbonic anhydrase, 76
Castration, 114
Cathepsin, 49
Caudal gland, 15, 88, 100
Cellulase, 48, 50
Cereals, 125, 153
Cerebral ganglion, 19, 95, 106, *110*, 113, 114, 115
Cerebral gland, 95, 113
Chemical control, 133, 144, 146–50, 153–4
Chitinase, 49, 50
Collar cells, *103*, 115
Collostyle, *40*
Colour, 15, 97
Colour inheritance, 97
Common duct, 17, 85, 86, 91, 92, 114
Competition for resources, 130, 135
Contact poisons, 147, 148
Control measures for slugs, 138, 142–54
 integration of, 151–4
 timing of, 152–4
Controlling factors, 129–35
Copper, 70, 75, 77
 aceto-arsenite (paris green), 149
 oxychloride, 150
 sulphate, 147, 148
Copulation, 87, 88–91, 92, 97, Fig. 41, 100, 108
Corvus corone (crow), 134
 frugilegus (rook), 134
 monedula (jackdaw), 134
Courtship, 19, 87, **88–91**
Creeping buttercup (*Ranunculus repens*), 37, 38
Crop, *16*, *45*, 47–50, 52, 54, 56, 57
 juice, 47, 48–9, 50, 52, 56, 57
Crop husbandry, 151, 153
Cropping systems, 125
Cross fertilisation, 92, 97
Cryptochiton, 57
Cultivation, 125, 133, 140, 143, 144, 153–4
Cultural control, 143–6
Cytopempsis, 82

DART, 19, 28
Dart sac, 19
Davainea proglottina, 136, 141
Defence mechanisms, 69
Density, estimation of, 116
Density dependent factors, 130, 134–5, 152

Density independent factors, 130, 131–4
Deroceras, 30
Development, 12, 94–6
 speed of, 29, 131, 132, 133, 135
Diaulic, 17
Dicrocoelium dendriticum, 136, 141
Digestion, 44–58, 59, 95
Digestive cells, 52, *53*, 55, 56
 glands, 16, *45*, 52–3, 54, 55, 57, 60, 61, 63, 70, 77, 80, 81, 95;
 enzymes, 52; openings, 50, *51*, 52
 system, *16*, 44, *45*, 62, 69, 95, 109; function, 54–8
Digitate ganglion, 102, *103*, 105
Dispersal, 126–7
Distribution, 109, 116, 117, 151
 geographic, 21, 23, 25, 26, 27, **28**, 30, 32, 33, 34, 35, 36
 horizontal, 125–6, 133
 spatial, 123–7, 135
 in time, 127–9, 135
 vertical, 123–4
Diurnal rhythm, 54, 79, 108
Dorsal bodies, 113
Dorsal food groove, 39, 42, 47
Downy mildew, 142
Drinking, 71
Drought, 120, 124, 125, 129, 135, 144

EARTHWORMS, 154
Ecology, 116–37
Economic importance, 138
Ectocone, 15
Eelworm, 139, 151
Egg, 19, 62, 85, 88, 91, *92*, 93, 94, 96, 97, 113, 118, 135, 153
 hatching, 94, 95, 96, 132
 laying, 85, 88, 92–4, 123, 131, 133
 shell layer, *92*
Embryo, 60, 85, 94
Endocone, 15
Endocrinology, 112–15
Endodontidae, 22, 24
Energy balance, 59
 cycle, 129
 reserve, 59
 shortage, 129
Entosiphonus rhomboideus, 142
Epiphallus, **27**, 90
Esterase, 55
Evaporation, 71, 72, 73, 108
Evolution of slugs, 9, 12, 21
Excreta, 16, 79, 101
Excretion, 70, **77–82**, 101

Excretory cells, 52, *53*, 55, 56, 80
  potential, 81
  system, 16, **77–9**
Eye, 11, 14, 95, **102**, *103*, 104

FAECES, 16, 43, 50, 53, 54–5, 56, 57, 80
Fasciola, 141
Fecundity, 131, 132, 133
Feeding, 37–44, 105–6
  activity, 54, 108, 122, 129, 132, 144, 145
  track, 39, Fig. 24
Female groove, *86*, 87, 91, 92
Fertilisation, 36, 91–2
  pocket, 92
Field studies, 123–35
Flies (sciomyzids), 134
Flooding, 119, 120, 121, 123
Flowers, 140
Food, 23, 25, 28, 30, 32, 33, 34, 35, **37–8**, 54, 60, 73, 85, 105, 133–4, 135, 142, 145, 150
  preferences, 37, 105–6, 144–5
  quality, 133
  reserve, 53, 59, 69
Foot, 15, 20, 60, 61, 63, 66, 94, 98–100, 101, 105, 109
  fringe, 15, 24
Formulation, 148
Frost, 120, 124, 125, 129, 130, 135
Fungi, 105, 134, 142

GALACTOGEN, 60, 85
α-Galactosidase, 48
β-Galactosidase, 48
Gametes, 17, 83, 84, 85
Gametogenesis, 83, 84, 85, 96, 115
Ganglia, 19, 95, 107, 109, 111–12
Gastric grooves, 50, *51*
Gelatinase, 49
Genetics, 97
Genital atrium, 88, 91
Geomalacus, 24
  maculosus, 24
Geotaxis, 106–7
Germinative cells, 83
Glial cells, 109
Glucose, 56, 57, 59, 60, 105
α-Glucosidase, 48
β-Glucosidase, 48
Glyceryl ethers, 61
Glycogen, 60, 70, 79, 109, 113
Gonad, 17, 62, 84, 85, 92, 96, 97, 114, 115
Grain-hollowing, 139

Grass, 125, 140, 144
Gravity perception, 106–7
Grazing, 139, 140
Growth, 37, 38, 59, 95, 96–7, 113, 114, 115, 132, 133
  curve, 96
Guanine, 80, 81
Gulls, 134

HABITATS, 125
Haemagglutinins, 69
Haemocoel, 9, 17, 65, 70, 72, 76, 82, 101
Haemocyanin, 70, 74, **75–7**, 82
Hair cells, 106
Haldane effect, 77
Hand-sorting, 120
Hazards, 154
Heart, 17, 67, 68, 69, 75, 76, 77, 94, 95
  beat, 67, 68
Hedgehogs, 134
Helicarionidae, 22
Helicidae, 22, 80
Helix, 10, 11, 44, 47, 55, 60, 63, 65, 67–8, 69, 72, 73, 75, 76, 77, 79, 80, 81, 82, 99, 101, 102, 105, 106–7, 113
Hemphillia, 27
Hepatic ducts, 52
  lobe, 94, *95*
Hermaphrodite duct, 17, *19*, 83–4, 85, 90, 91, 92
  gland (see also Gonad), 83, 85
Hesperarion, 27
Heterodera rostochiensis, 151
  schachtii, 151
Heterurethra, 17, 22
Home range, 126, 127
Hormones, 114–15
Horticultural crops, 133, 140
Humidity, 71–2, 94, 122, 132
  sensitivity, 107–8, 121, 129, 131, 132, 133
5-Hydroxytryptamine, 68, 112
Hyponotum, 15, 35
Hypoxanthine, 79

IMAGE PERCEPTION, 102
Immune reactions, 70
Incilaria, 28
Infantile phase, 96–7
Infection, 69
Infra-red sensitivity, 104
Innate capacity for increase, 131, 132
Insects, 134, 137, 151, 154
Integrated control, 151

Index

Protein, metabolism, 61, 64
Protozoa, 134, 136
Pseudaneita, 14, *34*
Puberty, 96
Pulmonary veins, 64
Pulmonata, 10
Pulmonate classification, 11, 22
Purines, 79–81, 82
Purinotelic, 79

RADULA, 9, 15, 22, **38–44**, Fig. 22, 57, 95
Radular gland, 15, *40*, 43
Rathousia, 36
Rathousiidae, 22, 36
Rectal caecum, *16*, *45*, 54
Rectum, 14, *16*, *45*, **54**, 55, 77
Reference works, 22–3
Regeneration, 84, 97, 102, 106
Renopericardial duct, 17, 81
Reproduction, **83–94**, 108, 114
Reproductive capacity, 125
 maturation, 59, 93, 96, 114, 132
 system, 17, *19*, 62, 69, *86*, *87*, 96, 109
Respiration, 59, **62–4**, 65, 94
Respiratory epithelium, 17, 64
 quotient, 64
 rate, 62–4
Riccardoella limacum, 134–5
Root crops, 125

SALIVA, 42, 46, 57
Salivary ducts, 46
 glands, 15, **46**, 52
Salt and Hollick process, 118, 123
Sampling methods, 116–23, 129
 units, 117
Sarcobelum, 19, 30, 87, 88
Schistocera gregaria, 151
Schwann cells, 110
Screening methods, 147, 148
Searching, 121, 123
Seasonal fluctuations in numbers, 127–9
Seed dressings, 150, 154
Self-fertilisation, 91–2, 97
Seminal vesicle, 85
Semper's organ, 113
Senile phase, 96
Sensory structures, 102–9
Sheath cells, 109
Shell, 9, 10, 11–12, 23, 24, 29, 30, 34, 80, 94, 134,
 gland, 94

Shelter, 108, 125, 126, 127, 130, 133, 135, 144, 146
Shredding, 139, Fig. 57
Sigmurethra, **17**, 22
Size, 12, 37–8, 41, 43, 62
Skin, 9, 14, 71, 72, 100–1, 112
 respiration, 65
Slit seeding, 140
Slug characteristics, 11
 classification, 21–36
 damage, 138, 139, 145, 151–2
 form, 9, 11
Smell, 102, 105–6
Sodium pentachlorophenate, 147, 148
Soil, 125, 126, 144, 146, 152, 153, 154
 compaction, 133, 144
 moisture, 93, 132
Soluble food uptake, 56–7
Speed, 100
Sperm, 17, 83, 84, 85, 88, 90, 91, 92, 96, 114–15
 mass, 88, 89, 90, 91–2
Spermatheca, *86*, *87*, 88, 91, 101
Spermathecal duct, *86*, 88, 91–2
Spermatophores, 19, 25–6, 29, 88, *90*, 91
Spiral cleavage, 94
Sprays, 148
Statocyst, 94, **106–7**
Statoliths, 106
Steroids, 61–2, 115
Stinging nettle (*Urtica dioca*), 37–8
Stomach, *16*, **50**, *51*, 52, 54, 58
 poisons, 147, 148
Streamlining, 12
Sturnus vulgaris (starling), 134
Stylommatophora, 11, 16–17, 22
Sugar beet, 140, 153
Suprapedal groove, 15, *18*
Synapse, 110–11
Syngamus trachea, 141
Systellommatophora, 11, 14, 17, 22

TASTE, 102, 105, 145
Teeth, 15, 23, 39, 41, *42*, 43
 lateral, 16, 41, *42*
 marginal, 16, 41, *42*
 rachidian (central), 15, 41, *42*
 replacement, 42–4
Temperature, 43, 62, 63, 64, 65, 84, 93, 104, 120–1, 122, 132, 133, 140, 146, 153
 acclimation, 62–3
 acclimatisation, 62
 adaptation, 62
 sensitivity, **107**, 108

Tentacles, 11, *13*, 20, 23, 41, **102–6**, 108, 115
Testacella, 12, 15, 23, 42
  haliotidae, 23
  maugei, *13*, 23
  scutulum, 23, Fig. 10, *42*
Testacellidae, 22, 23
Tetrahymena rostrata, 69, 136
Thin cells, 52, *53*
Thorotrast, 54–5, 57, 69
Tillering, 140
Tipula, 149
Tissue defence mechanisms, 69
Torsion, 10, 23, 95
Toxicity, 146–8, 154
Tracheopulmonata, 17
Transplants, 114
Trapping, 120–1, 126, 129
Trehalase, 48
Trematodes, 134, 136
Trifid appendage, *86*, 87, 89
Triglycerides, 61
Trituration, 47, 56
Trophosphongium, 109
Typhlosole, 50, *51*

ULTRAFILTRATION, 82
Urea, 80, 81
Urease, 80
Ureter, 14, 16–17, 77, *78*, 79, 81–2
Uric acid, 79–82
Uricotelic, 79
Uridine diphosphates, 60
Urine, 82

Vaginulus, *13*, 19, 84, 93
Vaginulus borellianus, 35–6, 83, *87*, 93, 102, 115

Varietal susceptibility, 145
Vas deferens, *86*, *87*, 90, 91
Vegetation, 125, 146, 151–2
Veins, 64, 67, 75
Ventricle, 17, 67
Veronicella floridana, 36
  moreleti, *13*
Veronicellidae, 13, 15, 17, 22, 35, 36, 87
Visceral ganglion, 19, *110*
Visual pigment, 102
Vitamins, 38, 102

WASHING, 118, 123
Water content, 71–3
Water-holding capacity, 132
Water loss, **71–3**, 94, 108, 119
Water relations, 33, **71–3**, 93–4
Weather, 120–1, 127, 130, 131, 132, 133, 135, 140, 145, 146, 149, 150, 151, 152
Weight cycle, 71, 73
Wheat, 105, 122, 138, 139–40, Fig. 56, Fig. 57, 144, 146, 150, 152, 154
  bulb fly, 146, 150
Wind, effect of, 72, 73, 106, 108
  sensitivity, 108

XANTHINE, 79–80, 81
Xylanase, 48
X-rays, 43, 47, 54–5, 84, 96–7

Zacoleus, 27
  idahoensis, 27
Zygote, 19, 85, 92, 94

LIBRARY